Rescuing All Our Futures

Recent Titles in
Praeger Studies on the 21st Century

Vitality and Renewal: A Manager's Guide for the 21st Century
Colin Hutchinson

The Foresight Principle: Cultural Recovery in the 21st Century
Richard A. Slaughter

The End of the Future: The Waning of the High-Tech World
Jean Gimpel

Small is Powerful: The Future as if People Really Mattered
John Papworth

Disintegrating Europe: The Twilight of the European Construction
Noriko Hama

The Future is Ours: Foreseeing, Managing and Creating the Future
Graham H. May

Changing Visions: Human Cognitive Maps: Past, Present, and Future
Ervin Laszlo, Robert Artigiani, Allan Combs, and Vilmos Csányi

Sustainable Global Communities in the Information Age: Visions from Futures Studies
Kaoru Yamaguchi, editor

Chaotics: An Agenda for Business and Society in the Twenty-First Century
Georges Anderla, Anthony Dunning, and Simon Forge

Beyond the Dependency Culture: People, Power and Responsibility in the 21st Century
James Robertson

The Evolutionary Outrider: The Impact of the Human Agent on Evolution
David Loye, editor

Culture: Beacon of the Future
D. Paul Schafer

Caring for Future Generations: Jewish, Christian, and Islamic Perspectives
Emmanuel Agius and Lionel Chircop, editors

Rescuing All Our Futures

The Future of Futures Studies

Edited by
Ziauddin Sardar

Praeger Studies on the 21st Century

Westport, Connecticut

Published in the United States and Canada by Praeger Publishers
88 Post Road West, Westport, CT 06881.
An imprint of Greenwood Publishing Group, Inc.
Printed in the United States of America

The paper used in this book complies with the
Permanent Paper Standard issued by the National
Information Standards Organization (Z39.48–1984).

10 9 8 7 6 5 4 3 2 1

English language edition, except the United States and Canada,
published by Adamantine Press Limited, Richmond Bridge House,
417–419 Richmond Road, Twickenham TW1 2EX, England.

First published in 1999

Library of Congress Cataloging-in-Publication Data

Rescuing all our futures : the future of futures studies / edited by
Ziauddin Sardar.
p. cm.—(Praeger studies on the 21st century, ISSN
1070–1850)
Includes bibliographical references and index.
ISBN 0–275–96558–9 (alk. paper).—ISBN 0–275–96559–7 (pbk. :
alk. paper)
1. Forecasting—Study and teaching. 2. Pluralism (Social
sciences) I. Sardar, Ziauddin. II. Series.
CB158.R48 1999
303.49′09′05—dc21 98–45366

Library of Congress Catalog Card Number: 98–45366

ISBN: 0–275–96558–9 Cloth
0–275–96559–7 Paperback

For Maha, Zaid and Zain
Atif, Nasif and Hana
Noori and Lubna

Who will dream different dreams and
make different futures possible

Contents

Contributors

Sean Cubitt is Reader in Video and Media Studies and Head of Screen Studies at Liverpool John Moores University. A member of the editorial boards of *Screen*, *Third Text* and the *International Journal of Cultural Studies*, he has published widely on contemporary media, arts and culture. He is the author of *Timeshift* (Routledge, 1991), *Videography* (Macmillan/St Martins Press, 1993) and *Digital Aesthetics* (Sage, 1998).

Email: s.cubitt@livjm.ac.uk

Merryl Wyn Davies is a writer, television producer and futurist. Having worked for the BBC for a decade, on such programmes as 'Everyman' and 'Heart of the Matter', she now functions as an independent producer and consultant. She is author of *Knowing One Another: Shaping an Islamic Anthropology* (Mansell, 1988), co-author of *Barbaric Others: A Manifesto on Western Racism* (Pluto, 1993) and co-editor of *Beyond Frontiers: Islam and Contemporary Needs* (Mansell, 1989). Forever Welsh, she lives and works in Kuala Lumpur, Malaysia.

Email: mwd@pc.jaring.my

Steve Fuller is Professor of Sociology and Social Policy at the University of Durham, UK. He pioneered the research programme on 'social epistemology' and edited the journal *Social Epistemology* for several years. His books include *Social Epistemology* (Indiana University Press, 1988), *Philosophy of Science and Its Discontents* (2nd edn., Guilford Press, 1993), *Philosophy, Rhetoric and the End of Knowledge* (University of Wisconsin Press, 1993), *Science* (Open University Press, 1997), *Being There with Thomas Kuhn* (University of Chicago Press, 1999) and *The Governance of Science* (Open University Press, 1999).

Email: steve.fuller@durham.ac.uk

Ted Fuller is the director of the Foresight Research Centre and senior member of staff at Durham University Business School. His research interests include complexity, business foresight, enterprise and economic development, and information and communications technologies. His career has included the ownership of a small manufacturing business, business directorships, lecturing, consultancy and public service. He has published many academic and practitioner texts and software in the small business field. Most current research in the Centre is related to small enterprises and entrepreneurship for stakeholders, who include government, the corporate sector, the academic community, futurists, local communities and entrepreneurs.

Email: ted.fuller@durham.ac.uk

Susantha Goonatilake is at the New School for Social Research, New York. He is an internationally renowned scholar who comes from two disciplinary backgrounds, electrical engineering and the sociology of science and technology. His publications include *Aborted Discovery: Science and Creativity in the Third World* (Zed, 1984), *Crippled Minds: An Exploration into Colonial Culture* (Vikas, 1982), *Merged Evolution: The Long Term Implications of Information Technology and Biotechnology* (Gordon & Breach, 1998), and *Toward Global Science: Mining Civilizational Knowledge* (Indiana University Press, 1998). He was on the advisory board of *The Encyclopaedia of the History of Science, Technology and Medicine in Non-Western Cultures* (Kluwer, 1997).

Email: susanthag@hotmail.com

Sohail Inayatullah is a political scientist at the Communication Centre, Queensland University of Technology. He is on the editorial boards of *Futures* and *Futures Studies*. He is also the associate editor of *New Renaissance* and, since 1990, an executive board member of the World Futures Studies Federation. He has been active in the futures field since 1981. He has authored/edited a number of books and nearly 200 papers, articles and magazine pieces. His most recent publication is *Futures Studies: Methods, Emerging Issues and Civilisational Visions*, a multimedia CD-ROM reader. Forthcoming is *The University in Transformation: Global Perspectives on the Futures of the University*.

Email: s.inayatullah@qut.edu.au

Anne Jenkins is a Research Associate at the Foresight Research Centre, Durham University Business School, UK. Her research and publications are focused on small business futures, grounded in experience of many aspects of small business research including international trade, information management, supply chains and IT. She is currently testing an environmental scanning system for identifying and analysing trends. She has worked as a management consultant for multinational companies and spent several years working in small businesses. Her Ph.D. work focuses on developing methods of sensing and exploring change for futures studies.

Email: Anne.Jenkins@durham.ac.uk

Kirk W. Junker is a Lecturer at the Centre for Science Education at the Open University in Milton Keynes, UK. Before joining the Faculty of Science at the Open University, he was an environmental litigation attorney in the USA at the Pennsylvania Department of Environmental Protection, and taught environmental law at Duquesne University. He publishes in the fields of law, environmental law, science studies and communication. He is the editor of *Science and the Public* (Open University Press, 1998), co-editor of *Communicating Science: Professional Contexts* (Routledge, 1999), and index editor for the *International Encyclopaedia of Comparative Law* (J.C.B. Mohr).

Email: k.w.junker@open.ac.uk

Vinay Lal is Assistant Professor of History, University of California, Los Angeles. He was the William R. Kenan Fellow in the Society of Fellows in the Humanities at Columbia University for 1992–3, and has held fellowships from the National Endowment for the Humanities and the American Institute of Indian Studies. He has also been associated as Visiting Fellow with the Committee for Cultural Choices and Global Futures in New Delhi. His teaching and research interests are very wide, and his papers, articles and reviews, on subjects as diverse as the Indian diaspora, colonial Indian history, popular Indian cinema and culture, the politics of sexuality, human rights, and the global American hegemony, have appeared in over three dozen periodicals. He is the author of *South Asian Cultural Studies: A Bibliography* (Manohar, 1996) and editor of *Plural Worlds, Multiple Selves: Ashis Nandy and the Post-Columbian Future*, a special double issue of the journal *Emergences* (1996–7).
Email: vlal@history.ucla.edu

Eleonora B. Masini is Professor of Social Forecasting at the Faculty of Social Sciences, the Pontifical Gregorian University, Rome. A founder and former President of the World Futures Studies Federation, and former chair of the Futures Research Committee of the International Sociological Association, she co-ordinated the UNESCO 'Futures of Cultures' project. She is the author of *Why Futures Studies?* (Grey Seal, 1993) and Vice President of the Aurelio Peccei Foundation and member of the Club of Rome.
Email: fmasini@pelagus.it

Graham H. May is Principal Lecturer in Futures Research in the School of Built Environment at Leeds Metropolitan University and Convenor of the UK Futures Group. He is Course Leader of the Masters degree in foresight and futures studies, the only such course in the UK, and has more than ten years' experience teaching futures at undergraduate and post-graduate levels. His publications include *The Future is Ours: Foreseeing, Managing and Creating the Future* (Adamantine Press, 1996), chapters in *Tomorrow's Company: Sustainability and the Future* and *The Education Yearbook 1998* and host of papers and articles in such journals as *Futures* and *Futures Research Quarterly*. With qualifications in geography and urban and regional planning, he is a member of the Construction Panel of the UK Government's Foresight Programme.
Email: G.May@lmu.ac.uk

Ivana Milojevic currently works as a social researcher in the area of mental health. Her academic career includes three years as teaching assistant at the University of Novi Sad in Yugoslavia. Her education and research interests are in the area of women's studies, sociology and futures studies. She has published several papers on issues dealing with gender and the future, including 'Learning from Feminist Futures', in David Hicks and Rick Slaughter (eds.), *1998 World Yearbook For Education* (Kogan Page, 1997) and 'Towards a Knowledge Base for Feminist Futures

Studies', in Rick Slaughter (ed.), *The Knowledge Base of Futures Studies*, Vol. 3 (Futures Study Centre/DDM, Melbourne, 1996).

Email: Ivana.Milojevic@panon.ns.ac.yu

Ashis Nandy, one of the most respected and influential of Indian writers, is a psychologist, futurist and the author of *The Intimate Enemy* (OUP, 1983), *Traditions, Tyranny and Utopias* (OUP, 1987), *The Tao of Cricket* (Penguin, 1990), *The Illegitimacy of Nationalism* (OUP, 1994), *The Savage Freud* (OUP, 1995) and other books. He is Chairperson of the Committee for Cultural Choices and Global Futures and Senior Fellow and former Director of the Centre for the Study of Developing Societies.

Email: nandy@unv.ernet.in

Jan Nederveen Pieterse is Associate Professor in Sociology at the Institute of Social Studies in The Hague. He has taught at universities in Ghana and the United States and has been visiting professor in Indonesia and Japan. He is co-editor of the *Review of International Political Economy* and advisory editor of the *European Journal of Social Theory*. His books include *White on Black: Images of Africa and Blacks in Western Popular Culture* (Yale, 1992), *Empire and Emancipation* (Praeger, 1989), *Christianity and Hegemony* (Berg, 1992), *Emancipations, Modern and Postmodern* (Sage, 1992), (with Bhikhu Parekh) *The Decolonization of Imagination*, (Zed, 1995; OUP, 1997) and *World Orders in the Making: Humanitarian Intervention and Beyond* (Macmillan, 1998). He is currently working on *Development—Deconstructions/Reconstructions* for Sage.

Email: nederveen@iss.nl

Jerome Ravetz is a 'philosopher at large' concerned with science in the broadest sense. Former Reader in the History and Philosophy of Science at the University of Leeds, he is author of the seminal work *Scientific Knowledge and Its Social Problems* (OUP, 1971), which has recently been republished (Transaction Publishers, 1996). His other books include *The Merger of Knowledge with Power* (Mansell, 1990), (with Silvio Funtowicz) *Uncertainty and Quality in Science for Policy* (Kluwer, 1990) and (with Ziauddin Sardar) *Cyberfutures: Politics and Economy on the Information Superhighway* (Pluto Press and New York University Press, 1996; Shi YuShi, 1997). He has pioneered the concept of 'Post-Normal Science'. He is based in London, as Director of Research Methods Consultancy Ltd.

Email: jrravetz@compuserve.com

Ziauddin Sardar, one of the most noted intellectuals of the Muslim world, is a writer, futurist and cultural critic. He has published some thirty books and over two hundred papers, essays and articles. His most recent books include *Postmodernism and the Other* (Pluto Press, 1998), *Cultural Studies for Beginners* (Icon Books, 1997), *Chaos for Beginners* (Icon Books, 1998), *Orientalism* (Open University Press and University of Minnesota Press, 1999) and, as co-editor, *Cyberfutures: Politics and Economy on the Information Superhighway* (Pluto Press and New York University Press, 1996). He is a Consulting Editor of *Futures* and Visiting Professor of Science Policy at Middlesex University.

Email: zsardar@compuserve.com

Richard A. Slaughter is Director of the Futures Study Centre in Melbourne and a consulting futurist who has worked with a wide range of organisations in many countries and at all educational levels. Slaughter is an internationally renowned futurist, a Fellow and member of the executive board of the World Futures Studies Federation (WFSF) and a professional member of the World Future Society. He is Consulting Editor of *Futures* and series editor of the Routledge series on 'Futures and Education', board member of the *Journal of Futures Studies* and series editor of *The Knowledge Base of Futures Studies* (FSC, Melbourne). His publications include *Education for the 21st Century* (Routledge, 1993), *The Foresight Principle—Cultural Recovery in the 21st Century* (Adamantine/Praeger, 1995), *New Thinking for a New Millennium* (Routledge, 1996) and, with David Hicks, the *World Yearbook of Education 1998: Futures Education* (Kogan Page, London 1998). He has also published a series of futures resource books and is editor of the monthly *Australian Business Network Report.*

Email: fsc@alexia.net.au
Web site: http://www.futures.austbus.com

S. P. Udayakumar is a Research Associate and Co-Director of Programs at the Institute on Race and Poverty, University of Minnesota–Minneapolis/Saint Paul. He has published numerous papers and essays on futures studies. An active member of the World Futures Studies Federation and the International Peace Research Association, Udayakumar runs, together with his wife, Meera, the South Asian Community Centre for Education and Research (SACCER) Trust in Nagercoil, Tamil Nadu, India.

Email: spkumar@tc.umn.edu

Morgen Witzel taught history at the University of Victoria in Canada before joining the research faculty of London Business School in the UK. He remains a visiting lecturer at LBS, co-teaching an MBA elective course on business in Greater China. In 1990 he helped found Western Writers' Block, an editorial and publishing services business, in which he is senior partner. Among his recent and forthcoming publications are the *Dictionary of Business and Management* (ITBP, forthcoming); 'General Management' (with Malcolm Warner) and 'Business Information' (with Anne Jenkins) in the *International Encyclopaedia of Business and Management* (ITBP, 1996); biographies of Charles Handy, Luther Gulick, Marshall McLuhan and Tom Peters for the *Handbook of Management Thinking* (ITBP, 1998); 'General Management: Back to the Future?' (with Malcolm Warner) in *Human Systems*

Management (1988) and 'General Management Revisited' (with Malcolm Warner) in the *Journal of General Management* (forthcoming). He is the editor of a series of business history reprints from Thoemmes Press, and writes a bi-monthly series of articles on 'Managing before Management' for the *FT Management Review*.

Email: morgen@wwblock.demon.co.uk

Rescuing All Our Futures

Introduction

ZIAUDDIN SARDAR

The origins of this book lie in a paper that was commissioned by Unesco. It was to appear in a special issue, devoted to 'Exploring the Future', of the *International Journal of Social Sciences.*[1] Entitled 'Colonising the future: the Other dimension of futures studies', the paper was deemed 'too sensitive' and subversive. The editors rejected the paper as it upset not just the referees but all those who read it. It subsequently appeared in a much more august journal: *Futures.*[2] Not surprisingly, it generated a lively controversy and equally vigorous responses and rejoinders.[3] Subsequently, the Indian journal *Seminar* picked up the theme of the paper and devoted a whole issue to exploring the subject.[4] This book not only brings together the futurists and scholars who contributed to the debate in *Futures* and *Seminar* but widens the contours of the whole argument.

The argument itself is simple but compelling: the future is being colonised and futures studies has become an instrument in that colonisation. 'Anticipating the future nowadays', as the opening essay states, 'means little more than forecasting the future' and 'the future is little more than the transformation of society by new Western technologies'. The future is thus locked into a single, dominant but myopic projection.

In this scenario, technology is seen as the sole resource of the West; a resource that will shape *the* future for us all. But that future will be firmly in the image of the West—the non-West will simply be an appendage to that future. Through its obsession with forecasting and prediction, its technology fetish, and its almost total neglect of non-Western cultures and visions, futures studies promotes the colonisation of the future. Indeed, futures studies has become synonymous with Western domination. Moreover, futures studies, like cultural studies, evolved as an intellectual movement concerned with changing the (inevitable) future, but it is now, also like cultural studies, becoming a discipline with fixed contours and boundaries. It is thus in danger of becoming an instrument for foreclosing the future, particularly non-Western futures.

This thesis is not original. In the late 1970s, Krishan Kumar made similar observations in *Prophesy and Progress.*[5] In the early 1980s, Eleonora Masini, then President of the World Futures Studies Federation, articulated concerns about an absence of visions and pluralism in futures studies.[6] Since then a number of futurists, both from the West and the non-West, have expressed anxiety about the domination of deterministic Western modes of thought and action in the futures field. The overall concern with the dominance of a single mode of expression in future studies has now reached a critical mass. This book specifically sets out to examine the issues involved: how futures are foreclosed to a single, extrapolation-of-the-present scenario; how alternative and pluralistic futures can be opened and explored; how futures studies can be made pluralistic; and how non-Western worldviews, issues and concerns can be made an integral part of futures studies. In other words, *Rescuing All Our Futures* attempts to change the political and intellectual landscape of futures studies.

As Masini, who is perhaps best described as the *grande dame* of futures studies, states in her classic paper, reproduced here with only minor changes, 'there is not one future, there are many futures'. To ensure that plurality is maintained, she insists that visions, and the whole process of visioning, should become the core of futures studies and research. Visions focus on desirables and usher genuine, pluralistic change by nurturing the real 'seeds of change'. These seeds of change, Masini argues, are almost always to be found on the periphery: in the non-Western cultures, among women and children, and 'outsiders' such as poets, artists and philosophers. The social responsibility of futurists is not just to aspire to transform the present by visions of alternative futures but to turn this undertaking into a project. Futures thinking, she argues, 'must be linked to social responsibility and to ethical values that are clearly expressed and defined'; and it should be a continuous learning process. The project of the futures discourse is then to persuade the world to move towards pluralistic futures though creative and imaginative analysis and productive action.

How does the West retain its firm control on the future? Sohail Inayatullah suggests that four types of power are involved: definitional (what is nature, truth, beauty and reality?), temporal (linear, developmentalist views of time), spatial (secular and urbanised) and economic (creating strong centre–periphery distinctions). S. P. Udayakumar argues that the 'three-in-one' product of nation-statism, scientism and developmentalism has become axiomatic for futures studies and ensures that the future continues on its present trajectory. He shows how the future is conceived, manipulated and managed by Western 'futures facilitators' and examines some of the classic works in this field. It does appear a bit sad that such a celebrated futurist as Dennis Gabor could actually suggest that a group of 'very smart people' should be asked to prepare a list of possible futures, but this kind of elitism has a long history in Western thought going back to Plato and his *Republic*.

Kirk Junker identifies the very ideas of prediction and forecasting as the main culprits—they work against multiple futures and for a single, linear future. Any notion of the future, notes Junker, is a notion that creates meaning. 'A picture of the future is a picture with meaning; an image of the future is an image with meaning.' Predictions thus become prescriptions; and images of the future become graphic prescriptions. There is thus a continuum of dominance: prediction–prescription–image. Junker shows how this works with the example of cloning. Even though there are serious moral problems with human cloning, the very prediction of cloning determines its future: 'predictions about the future of cloning in fact can create the future of cloning'. Over and above moral problems is the image of technology that will ultimately lead to human cloning. Once this image is in place, stories emerge that give meaning to this image of human cloning. The future vessel is filled with moral images that provide justification for the determined future: the infertile heterosexual married couple desperate for a child, the child suffering from a genetic disorder, the consumers queuing for designer babies. Thus the prediction–prescription–image continuum pulls the described future into the present.

Udayakumar suggests that globalisation has actually become a new salvation theology for a future world. Through globalisation Marx's vision of the high-tech business homogenising the planet has really come true—the irony that this vision arrives after the whole Bolshevik balloon had burst notwithstanding! But Jan Nederveen Pieterse sees globalisation as a way forward. Pieterse argues that the colonisation of the future is a routine process, a natural outcome of the 'colonisation of the life world'. Futures are becoming important, he argues, because ideologies have lost their impact; even neo-liberalism has 'probably passed its peak'. The much-trumpeted notions of 'global development' and 'global governance' are becoming quite meaningless. We need to move towards truly global futures because the idea of national citizenship is coming under strain and 'citizenship is becoming increasingly national *and* local, regional, global at the same time'. This does not mean that citizenship has simply become international. Pieterse would like to see the evolution of 'multicentric' global futures through the development of global public opinion. In this exercise both the ends and the means of shaping preferable global futures must be democratic and progressive. The route towards democratic global futures should be based, Pieterse states, on: '1. A stock-taking of socially progressive best practices and future proposals. This refers to innovation not simply in a technological sense, but in social practices, institutions, values and expectations. 2. Articulating them, stitching them together across domains and geographies. 3. The development of global public opinion concerning planetary predicaments and future options. 4. The development of multidimensional relations of negotiation, i.e. across different sectors and dimensions'.

But globalisation can generate conflict with local cultures. Anne Jenkins and Morgen Witzel argue that globalism and localism, which is about the survival of local identity and local communities, 'have a lock on both the present and the future—reducing both to a single denominator'. They suggest that futures studies should draw from history and work with multiple paradigms. 'There is a great deal of common ground between history as a discipline and futures studies: they use the same techniques, have the same responses to uncertainty, and, indeed, fight the same battles for acceptance in a society that seems determined to ignore, or at least unwilling to confront both the diversities of our pasts and the pluralities of our futures'. Just as history teaches us about diversities, about our different pasts, futures studies should emphasise our different futures. Just as historians use civilisations as a macro unit of analysis, so futures studies should focus on civilisational futures.

However, working with civilisations means appreciating non-Western notions of time—a point also emphasised by Inayatullah and Merryl Wyn Davies. Inayatullah argues that deeper, more challenging alternatives to Western notions of technocratic time can be found in the work of macrohistorians and critical traditionalists concerned about civilisational visions of the future—such as Islamic, Tantric, Maori and Hawaiian perspectives. 'These alternatives are generally long term and cyclical in their view of the future; concerned with future generations instead of an abstract future; collective in the sense that they involve family and community; spiritual; civilisational; committed to an alternative co-operative view of economics; concerned with leadership-creation (instead of merely efficiency in organisation); pluralistic in their strategy; and have roots in non-negotiable axioms'. Davies suggest that futurists concerned with pluralism have little alternative but to familiarise themselves with canons of non-Western civilisations; and she provides a short list of non-Western classics that should be on every futurist's bookshelf.

Futurists could, quite easily, start with looking at their language and attempting to make it less sexist. Ivana Milojevic shows how sexism is actually built in in the very terminology of futures studies. Like Jenkins and Witzel, she argues that women are being left out of futures studies, but goes on to show how their work is often undervalued and demeaned. This was not always so. In classical mythologies, the gods of the future were always women. But when gods became male 'so did time, so did the future'. Today, predictive forecasts and future trends are represented on graphs in the same way we draw the symbol for Mars, the god of war: an arrow pointing upwards, a phallic symbol. Is it surprising that the most popular symbol of the future is the 'Terminator'? But Milojevic does not want to return to some mythological goddesses-based futures—for they were a product of another time and another history. She wants to go forward to a futures studies that uses 'feminine' guiding principles and 'emphasize futures of education, parenting, community, relationships and health—the real grand issues'.

All these suggestions look outside futures studies and seek to transform the discipline by using external frames of reference. But is there enough scope *within* futures studies for multiculturalism? Richard Slaughter, Ted Fuller and Graham May suggest that futures studies can be transformed from within; it has all the resources and all we need is to activate and utilise them properly. Slaughter argues that opening up futures studies to pluralistic concerns means establishing it on the social level. He offers four distinct layers, or levels, through which this can be achieved. First is the natural capacity of the human brain–mind system to envisage a range of futures. Second is the clarifying, enlivening and motivating role of futures concepts and ideas. Third are analytic gains provided by futures tools and methods. Fourth are a range of practical and intellectual applications, or contexts. When all of these levels function in a co-ordinated way, argues Slaughter, grounds for the emergence of futures studies at the social level can clearly be seen; and as social discipline, futures studies would naturally be a multicultural and pluralistic one.

Slaughter actually illustrates his argument by providing a vision of a desirable future that would, arguably, be within reach if futures studies was to progress along such a path from individual to social capacity.

Ted Fuller, on the other hand, sees futures studies as a multilayered exercise in thinking concerned with a number of domains. In each domain, futures studies asks a pertinent question. In the philosophical domain, for example, futures studies becomes the question 'So what does this or that philosophical principle do to help us contemplate what is yet to be experienced?' When exploring political, economic, social and cultural futures the question for futures studies becomes 'How can the "theories" and meanings constructed within this domain more closely explain observations and guide the understanding and the enacting of progress?' As worldview, futures studies asks the question 'In what ways can the problem be viewed?' When focused on future generations, futures studies inquires, 'What are the consequences of today's action?' In forecasting, the question becomes 'What is the pattern of regularities and what is the limit of the context of those regularities?' As critical alternative thinking, futures studies becomes an assertion of preferred futures or a warning of impending futures. Provided we ask the right question, Fuller seems to argue, we can ensure plurality and pluralistic visions are innately ingrained in futures studies. While acknowledging that if futures studies was colonised by a single worldview it would become a 'fragmented specialism' and fail, Fuller asserts that 'the issue of plurality in thinking about the future barely need arise; such is the natural plurality of foresight'.

Graham May places his faith firmly in the complexity of modern times, which will ensure that futures are always complex and full of surprises. Empirical knowledge does have its place in futures studies, he argues, but it is not enough; 'other forms of knowledge, such as belief, faith, ethical and moral conviction and the envisioning of future images become equally important in creating the future'. The future, as well as the past and present, is not singular only in the English language; other cultures too have subscribed to these singular notions. But if we acknowledge, as we are now beginning to do, that there were several histories, that there are numerous presents, and accept the idea that human choice is involved at every level, then the potential for different futures to occur will be realised. This can best be done, May suggests, echoing Masini's argument, by treating futures as a long learning process. This involves a number of 'necessary characteristics' including 'open-mindedness, diversity, flexibility, creativity, accepting there will be few final solutions, continually seeking information, giving up control to maintain influence, welcoming ideas invented elsewhere, accepting a participatory approach, and regarding strategies as learning devices'.

But to those working in other disciplines, such as sociology of knowledge or the new-media arts, the problem of futures studies looks rather different. Steve Fuller rightly accuses futures studies of a 'relative lack of attention to the sense of historicity that invariably accompanies any judgement of what the future may hold'. He asks how one must be situated in history in order for the future to appear as it does. Fuller locates the linear projection of the present onto the past in what most people have accepted as the 'logic of natural science'. Seen from this perspective, 'science puts an end to history: once the natural trajectory of science is appropriately harnessed to the future of one society, history then simply consists of the rest of the world catching up by repeating the steps originally taken by that society'. The future of futures studies is tied firmly to the future of science. And Fuller presents his own, powerfully articulated, 'Republican vision' of science that is designed to keep the future of science, and hence all our futures, open and pluralistic.

If multicultural futures are a function of pluralistic notions of science, then they are just as intimately linked to our understanding of art and possible, alternative futures of art. In a very eloquent essay, Sean Cubitt argues that art can break the dominant hold of a single ideology on all our futures. 'Art', writes Cubitt, 'acts in many ways like the devotional sculptures, paintings and architecture of more spiritual times and cultures: to create a sense of a world beside, outside, beyond—a beyond which, for a secular age, is always future. It is the living practice of transforming the everyday into its negation.' But he is not concerned with just any art—he sees the new-media-based arts as providing a valuable opportunity to explore other ways; we can realise other futures through this exploration. Artists working in the new fields are creating new social and cultural networks, rebuilding existing machines, replacing the ideology of efficiency in communication with new modes of interpretation, including the right to be misinterpreted and misunderstood—in general, rejecting the 'Great White Future' and supplanting the controlling ethics without morality with a moral ethics. And because these artists are marginalised from the academy, they imbibe a true spirit of amateurism: 'the act of making for the love of it—a love which itself is a relation with the other and the future'. The new media arts thus 'provide us with an alternative model of futures studies: a practical, hands-on materialism in the age of the simulacrum'.

New-media arts, the notion of 'Republican science', scholarship from other disciplines, civilisational analysis, appreciating non-Western notions of time, learning from non-Western canons, moving towards genuine, pluralistic globalisations —all of these are essential for opening up futures studies to non-Western possibilities. But do they constitute something more than simple resistance? Why is the non-Western response to colonisation of the future, as Inayatullah laments, little more than imitation, nothing more than a fetish with Westernised futures? And is resistance by itself good enough. Surely Cubitt is right when he suggests that 'the business of building the future cannot be undertaken by placing yourself at every turn in relation to the dominant'. Resistance often leads the dominant worldview to a sense of its own importance and to the multiplication of efforts to dominate. There is also the need for producing genuine alternatives to the dominant modes of thinking and doing things.

Enter Susantha Goonatilake. He argues that the whole knowledge industry is set up in such a way that the non-West can respond only in terms of imitation. Neglect, lack of funding and an inferiority complex about indigenous knowledge do not only marginalise the non-Western academics and intellectuals from the mainstream; this amalgam also forces them to imitate their Western colleagues in a futile attempt to gain respect and credibility. Goonatilake argues for changes within the Western system of knowledge through interaction with the reservoir of knowledge from without: non-Western academics should explore their own cultures and use their indigenous knowledge to transform existing disciplines, thus performing a reverse colonisation. The interaction of cultural elements from within the non-West, including 'the local historical experiences and the conceptual elements that have emerged out of these experiences', with the dominant form of futures studies would, argues Goonatilake, be a productive encounter for both. It would creatively transform Western knowledge and bring non-Western knowledge to the fore. Goonatilake gives many examples from non-Western medicine, mathematics, philosophy and psychology where creative synthesis can be achieved and, in some cases, is already being achieved. Only through such encounters, such creative synthesis, can the non-West 'enter into meaningful and non-imitative discourses about the future'.

Goonatilake's argument does require one to have a certain degree of faith and confidence in Western knowledge structures and systems. It requires one to believe that Western knowledge systems will not simply cannibalise what they find of value in non-Western knowledge, that there would be a genuine synthesis rather than assimilation and consumption, and that, after they have been enriched, they would not continue on their current trajectories of domination. Vinay Lal has no such confidence. For him the whole structure of knowledge—from science to social science to humanities—is problematic from the non-Western perspective. Lal argues that Western disciplines such as anthropology, area studies and political science have in-built mechanisms for violence against the non-West. It is in the very nature of these disciplines to cannibalise, marginalise, assimilate, suppress, oppress and otherwise be violent towards non-Western societies. Through an analysis of the works of Samuel Huntington and Bernard Lewis, Lal shows that the non-West is offered only two stark choices: Western exceptionalism and Western universalism. It does not even have a choice of opting out, as dissent outside the Western framework is considered meaningless. Indeed, every act of dissent that calls into question the purportedly dissenting frameworks of knowledge thrown up by the West in recent years, whether encapsulated under the terms 'post-colonial' or 'postmodern', is construed as a retreat into romanticism, indigenism, nativism or tribalism. The future, therefore, cannot be opened up to plurality unless the dominant structures of knowledge themselves are decolonised. Lal sees the '20th-century university' as the first victim, closely followed by all variety of 'experts': only through reform—or is it removal?—of these two purveyors of knowledge can the future 'manifest itself as more ecological, multifarious and just'.

Are Lal and Goonatilake arguing against each other? I think their positions are complementary: undermining the Western academy and the cult of the Western expert, as well as interacting creatively with Western knowledge systems, requires non-Western thinkers and intellectuals to have confidence in their own reservoir of knowledge. This is what Goonatilake is arguing for; and Lal would like, although he does not clearly state this here, an alternative non-Western universe of knowledge and ideas pointing towards non-Western alternative futures. This requires evolution of new disciplines based on non-Western categories and concepts of knowing, being and doing. To some extent—and Goonatilake alludes to this—this is already happening. In her survey of non-Western futures studies, Davies provides a signpost to the efforts of Indian and Muslim scholars who are attempting to construct just such an alternative universe of discourse and knowledge.

Ashis Nandy, who has pioneered not just the idea but the whole movement of dissent against Western structures of knowledge, is a key player in this project. He makes it a social responsibility not just of all futurists but of the citizens of the globe to 'subvert the "inevitable" in the future', which is, after all, 'only another name for tomorrow that dare not be anything other than a linear projection of yesterday'. Nandy would like us to go forward by shaping contemporary versions of the prophetic. The prophets were not astrologers and predictors of the future. They were dreamers—'they dared to dream those dreams that were latent in all of us'. Their voice echoed our inner voice and it is in this instant resonance that their genius is to be found. There is resonance too between Nandy and Jerry Ravetz. To prophesy, Ravetz writes, 'meant to speak as inspired; and the utterances, usually as warnings, would necessarily be largely in the future and conditional tenses'. He suggests the use of prophecy as a means for dialectical thinking anchored in non-violence. Both Ravetz and Nandy argue that futures studies should transform itself into a mode of 'discourse more appropriate to the function of warning and awakening'. *Rescuing All Our Futures* sets out to do just that.

Notes

1. *International Journal of Social Sciences*, 45(137), 1993.
2. 'Colonising the Future: The "Other" Dimension of Future Studies', *Futures*, 25(2), March 1993.
3. See, for example, Richard A. Slaughter, 'Responding to the Inner Voices of an Irritated Text: A Response to "Colonising the Future"', *Futures*, 25(2): 188–90, March 1993; Sohail Inayatullah, 'Colluding and Colliding with the Orientalists: A Response to Zia Sardar', *Futures*, 25(2): 190–5, March 1993; Susantha Goonatilake, 'The Freedom to Imagine', *Futures* 25(4): 456–62, May 1993; Jerry Ravetz, 'It Will Nearly Always Be so', *Futures* 25(4): 462–4, May 1993; and Ashis Nandy, 'Future Studies: Pluralizing Human Destiny', *Futures* 25(4): 462–4, May 1993.
4. 'Futures', *Seminar* (New Delhi), 460, December 1997.
5. Krishan Kumar, *Prophesy and Progress*, London: Penguin, 1978.
6. Eleonora Masini, 'Reconceptualizing Futures: A Need and a Hope', *World Future Society Bulletin*, November/December 1982: 1–8.

1
The Problem of Futures Studies

ZIAUDDIN SARDAR

It is simple. The future has been colonised. It is already an occupied territory whose liberation is the most pressing challenge for the peoples of the non-West if they are to inherit a future made in their own likeness.

Even though thinking about the future is a tricky and hazardous business, it has become big business. Tricky because our conventional way of thinking does not normally incorporate the future—it requires considerable conscious effort to imagine how the future may unfold, what anticipated and unexpected possibilities lurk on distant horizons. Hazardous because the probability of getting one's forecasts wrong is very high. But this has not stopped the business of forecasting from spreading like a global fire. Anticipating the future nowadays means little more than forecasting the future. And forecasting is one of the major tools by which the future is colonised. No matter how sophisticated the technique—and techniques are becoming more and more refined and complex—forecasting simply ends up by projecting the (selected) past and the (often-privileged) present on to a linear future.

Despite numerous failures, unfulfilled promises, and misplaced optimism about the ability of technology to usher in a more humane and sane future, technological trends dominate the business of forecasting. The future is little more than the transformation of society by new Western technologies. We are bombarded by this message constantly from a host of different directions. The advertisements on television and radio, in newspapers and magazines, for new models of computers, cars, mobile phones, digital and satellite consumer goods—all ask us to reflect on how new technologies will transform not just our social and cultural environments but the very idea of what it is to be human. 'The future', according to a mobile phone company, 'is Orange.' According to a car manufacturer, 'The future is Vauxhall Vectra.' And computer junkies sing in unison, 'The future is the World Wide Web.'

It is highly significant that the filler material, what appears between programmes broadcast by satellite all over Asia, relies heavily on supposedly informative vignettes wherein Western experts reiterate the message of high technology as the creative potential of the future. The subtext is that the future technologies are the resource of the West, which will enable the non-West to have a future; the future it will have is a clone of the Western future. If that seems empowering and inclusive, it is only an illusory surface seduction that obfuscates how that future is made.

Business and corporate books, available globally and at any airport, tell us how we will all be internetted, tuned in to hundred of channels, working from home, and generally living in a technological bliss in the decades to come. For example, John Naisbitt's *Megatrends*[1] tells us that global trends are moving us from industrial to information age, national to world economy, representative to participatory democracy (where on earth can you find even an inkling of a participatory democracy?), and from either/or to multiple logic. The future will thus be better all around and for everyone. (Richard Slaughter has shown that Naisbitt's megatrends are half truths that cannot be clearly discerned and the book itself amounts to little more than a brochure for liberal capitalism.[2]) Peter Drucker's *Managing for the Future*[3] advises corporate types to hang on to their culture as the future is already with us. No need to contemplate what could or would happen in the next decades as they have already collapsed on us. And in *Rethinking the Future*,[4] we are told how the new science of complexity will enable us to manage uncertainty and generate new methods for creating tomorrow's advantages and strategies for growth and reinvent the basis for competition. The future, therefore, will not be much different in at least one respect: corporations will continue to dominate and they will have new theories and tools to maintain their domination.

The messages of such works are translated into visual metaphors by global television programmes such as CNN's 'Future Watch' and BBC's 'Tomorrow's World'. The sheer repetition and the intellectual and visual power with which this message is hammered home have profound consequences for our future-consciousness. It is thus not surprising that the vast majority of humanity thinks of the future only in terms of advertisement clichés, corporate strategies and gee-whiz technological gadgets. But 'reality' at ground level also comes wrapped in this notation. Consider the most profound developments of recent years. The information and communication revolution owes everything to advancements in technology. The cloning of 'Dolly' the sheep is a wonder of bioengineering—a short step from cloning man and not too far from redefining the whole notion of humanity. The eradication of dreaded diseases like smallpox and the containment of AIDS with a cocktail of new drugs are achievements of technology-based medicine. Moreover, technology is providing choices where there were none. A host of new fertility treatments now enables barren women to have a much-wanted child (or two, three or more), even choose the baby's sex; and in the not-too-distant future, its physical qualities, features and character could also be selected. All these developments make someone genuinely happy—and hold the rest of us in awe, imprisoned in the glare of technological advancements.

The inherent problem of the information revolution, however, is that most information is recycled in new packaged forms that are rigorously selected. To make children computer literate is a worldwide aspiration—to do the best for the future generation. In the non-West, it is seen as an imperative. Yet this well-intentioned determination not to be left behind again becomes the prime means of foreclosing the future. The resources available, the learning programs students can run on their computers, are more dominated by the West's selective vision than any library, bookstore or school yet devised. To surf the net is to immerse one's self in the worldview and interests of white male American college students. The grand strategies being offered of cheap technology that instantly circumvents marginality and exclusion (though it is not cheap enough to include the poor) again foster the idea that the world of all the knowledge that matters can by brought direct to classroom and home in the non-West instantly. Indeed information can be delivered—but it will have less non-Western content, more seductive clone-making intent than was ever conceived possible in the headiest days of the development decades.

There is thus an in-built momentum that seems to take us towards a single, determined future. Technology is projected as an autonomous and desirable force. Its desirable products generate more desire; its second-order side-effects require more technology to solve them. There is thus a perpetual feedback loop. One need not be a technological determinist to appreciate the fact that this self-perpetuating momentum has locked us in a linear, one-dimensional trajectory that has actually foreclosed the future. This trajectory is in fact an arch-ideology and like all ideologies it is inverse to the truth. An illusion of accelerated movement is created to shroud the fact that we are at best standing still, if not actually retarding. Faster and faster cars are actually not taking us anywhere but straight to a gridlock. No wonder concerned citizens in the West are giving up their cars for bicycles. Faster, intelligent and perfect computers are not solving any problems but creating newer ones. Being connected is a substitute for being a real community. All biotechnological advances have nightmarish underbellies and generate ethical paradoxes that are almost impossible to solve.

The reality is that we have reached a technological plateau. The futuristic revolution turns out to have had very little conception of the future—witness the billions that will have to be spent because computer programs cannot recognise dates beyond the year 1999. The millennium was beyond the consciousness of those who wrote and manufactured more and more sophisticated computer programs; instant solutions foreclose the future more effectively than planned ones. The future has been made only by projecting instant technological answers, and that means pushing forward the desires of the powerful. New technologies may appear to be better, faster and more promising, but in reality they do not improve our lives, or deliver greater material benefits to most of humanity or make us happier. While the belief in the power of technology to rescue our future continues to gain more ground, it is, in fact, a dangerously obsolete ideology. The future is thus waiting to explode.

The future is being colonised by yet another force. Conventionally, this force was called 'Westernisation', but now it goes under the rubric of 'globalisation'. It may be naive to equate the former with the latter—but the end product is the same: the process that is transforming the world into the proverbial 'global village', rapidly shrinking distances, compressing space and time, is also shaping the world in the image of a single culture and civilisation.

Globalisation can be identified with (at least) two main elements. The first is the economic wave of liberalisation that began in the 1980s and achieved global proportions after the fall of communism. Markets are becoming free from all state constraints and capital can now move across borders with ease. Multinational corporations move from country to country in their quest for cheap labour and tax exemptions. Globalisation has meant that a single consumer product, such as a computer, may actually be made in segments in several different places and put together in yet another place. While the management remains in the industrialised world, many sectors of manufacturing industry are now being located offshore in developing countries, where corporations can take advantage of cheap labour, lower taxes and liberal labour-protection and environmental-protection regimes. On the whole, the manufacturing facilities play very little part in fuelling the economy of the developing countries. Global capital is now shifting from resource-based and market-seeking investment to spatial optimisation and absolute maximisation of profit opportunities. The end result is that the economies of most countries, both industrialised and developing, are becoming dominated by consumer and lifestyle choices, production is being replaced by consumption as the central economic activity, and privatisation is becoming the norm.

Secondly, the wide acceptance of liberal democracy across cultures from Eastern Europe to Africa is leading to a total embrace of Western culture. Even though the trend towards universalisation of Western culture is actively contested, it has become the dominant norm, encouraged and aided by Hollywood, television, satellite, pop music, fashion and global news networks such as CNN, News International and BBC World. Thus, globalisation maintains all the well-known patterns of Western economic and cultural imperialism and goes further. It promotes a dominant set of cultural practices and values, one vision of how life is to be lived, at the expense of all others. And it has serious practical consequences: not only does it erode non-Western local traditions and cultural practices but it kills non-Western future options. Once again, the future is locked into a single, linear projection.

Furthermore, the future is being colonised in the way futures studies itself is being shaped into a discipline, with fixed boundaries, a set of basic principles and assumptions and all the other trappings of a crystallised discipline: established authorities, designated areas of research and thought, learned and professional organisations, bibliographic tools and study guides. As yet, futures studies is not a fully fledged discipline, although it has acquired the trappings of a discipline.

Traditionally, futures studies—like cultural studies—developed as an intellectual and social movement emphasising the plurality of futures, with a particular accent on alternative futures. As futures studies was domesticated and institutionalised, particularly in the corridors of American and European corporations, that emphasis began to evaporate. In this organisational framework, futures studies is synonymous with Western interests. The aim here is to preserve a future landscape where technological advances can be employed to maintain the hegemony of the West.

A direct outcome of the corporatisation of future studies is the belief that America is the locus not just for futures studies but for the future itself. This is the basic assumption and prime principle of much of the futures thought that emanates from even the grass-roots American organisations such as the Washington-based World Future Society (WFS). The Society's monthly journal *The Futurist*, bi-monthly abstract journal *Futures Survey*, and the learned journal *Futures Research Quarterly* are tools consciously designed to create a professional discipline that, like anthropology and Orientalism, serves the interests of the dominant culture. *The Futurist* regularly offers a comatosed vision of how technology will make our life better in—perhaps I am being charitable here—an unconscious attempt to validate the most debilitating forms of technological consumerism. *Future Survey*, the most important bibliographic tool in futures studies, seems to be totally blind to anything relating to the future that does not yield some kind of dividend or early warning signal for multinational corporations. The textbooks produced by WFS, like Edward Cornish's *The Study of the Future*,[5] set the worldview of *The Futurist* and *Future Survey* in disciplinary stone.

On the whole, futures studies is sponsored by scholars who are not just totally divorced from any political and cultural movements, but are quite unaware of the fact that the future has anything to do with critical questions of power, history and politics. Indeed, of the numerous intellectual movements that have swept American social sciences since the 1970s, few could be described as so utterly shallow and xenophobic, so opportunistically unreflective towards the non-West and so ahistorical in their analysis. The recently published *Encyclopaedia of the Future*[6] sums up the whole argument. Brief entries for countries such as India and civilisations such as Islam are there only as a necessary evil. Other countries involved in serious future planning, such as Malaysia, are conspicuous by their absence. There is absolutely no awareness of the numerous non-Western notions of the future, time, being or knowing. The list of the 'one hundred must influential futurists' contains only one person from the non-West. Clearly, the future is a Western territory that has no place for the non-West.

This is not to say that there are no Western futurists who use non-Western philosophies and modes of knowing as the basis for constructing alternative visions of the future. But even here both the research and the vision are strictly enframed within the European tradition of humanism—a tradition that is totally enveloped in the secularist worldview. The end-products of their labour are often a grotesque parody of non-Western thought, philosophy and tradition. As such, even the 'new spirituality' and 'values' that these more aware futurists offer ultimately conform to the dictates of Western secularism. Hence, it is always the secular forms of Eastern mysticism—like Zen Buddhism—with which these futurists find sympathy. The vast corpus of non-secular non-Western traditions is almost totally ignored. There is also the added irony of the product of Western humanism borrowing 'traditional thought' from a non-Western culture and subsequently presenting the repackaged confection to the natives. At best, the appropriation of non-Western ideas and thought amounts to little more than a second-hand regurgitation of 'Eastern mysticism', as in the case, for example, of the 'small is beautiful' guru E. F. Schumacher. But whatever his standing in the West, as a mystic in the Eastern tradition, Schumacher is decidedly an infant: the non-West has greater minds and a long historical tradition to learn from. At worst, non-Western ideas are used in an opportunistic exercise to make dubious reputations, as is the case with Fritjof Capra.

Thus, even when futures studies is allegedly borrowing and incorporating non-Western thought into its framework, it remains rooted firmly in Western philosophical ideas. All the future alternatives are actually worked out within this single, dominating, philosophical outlook. Other cultures are there, at best, for decorative purposes, or worse, to be used to prop up a system of thought and action that is actually responsible for the present dire predicament of humanity.

Of course, there is nothing special in the way futures studies has developed and is evolving towards a discipline. It is following a well-trodden path laid out in history by anthropology and Oriental studies and in more modern times by development studies. It is worth noting that these disciplines are remarkably similar in how they approach the non-West. Operating within a linear teleology that makes Western civilisation the yardstick for measuring progress, norms and values, these disciplines have evolved by using non-Western cultures and societies to define themselves and to develop and grow. In following suit, futures studies has not only colonised the future, it is itself becoming an instrument for maintaining and enlarging that colonisation. Futures studies thus has an unsavoury underside that is—ultimately—much darker than those of mere anthropology or Orientalism.

Elsewhere, I have compared the evolution of futures studies with the unfolding of development studies:

> In that field, Western 'authorities' were first created by citation analysis, literature surveys and study guides and the boundaries of the discipline were pegged to the research interests of these 'authorities'. The textbooks produced by these authorities became the essential teaching instruments in the Third World; while the masters of the discipline went to the Third World as consultants and authors of national development plans. It is only a matter of time until the 'experts' (identified in the *Encyclopaedia of the Future* and other disciplinary texts) make their appearance in the Third World as consultants to set up university departments and long-range future plans. Already the signs are ominous. Just as the 'national development plans' of so many developing countries reflect little concern or respect for indigenous culture and local needs, so many of the national future plans reflect the concern and interests of Western futurists rather than hopes and aspiration of the local population. The priorities of such future studies as Malaysia's Vision 2020, China 2000, Mexico 2000 have been set not by local populations but by the US Global 2000 report.[7]

The colonisation of the future by these powerful forces means that the future ceases to be an arena of action. We are in the domains of a new kind of colonisation that goes beyond physical and mental occupation to the seizure of our being and hence total absorption. Modernity tolerated our existence as an appendage to Western civilisation. But the postmodern future is less tolerant—it will settle for nothing less than complete assimilation of all non-Western societies into Western civilisation. Modernity raised the question 'Can the non-Western societies actually survive the future with their sanity and cultures intact?' The colonised horizon of the future forces us to ask a new question: 'Can we survive as distinctive entities, as something different, Other than the Western civilisation?'

Given the myopic vision and one-dimensional logic with which Western civilisation pursues its goals, and given the lack of concern for the future among non-Western intellectuals and thinkers and the consequent (almost) total lack of future-consciousness in many non-Western societies, my contention is that the prognosis is grim—unless we start to think more concretely and imaginatively about the future; unless we transform the future into a site of both real and symbolic struggles, and hence change the future by opening it to non-Western possibilities and move from the future to a plethora of futures. Changing the future means both questioning and resisting the forces, and the values and canonical myths associated with them, that have colonised the future.

Surviving the future thus involves confronting the deterministic, Western future and altering the political and intellectual landscape of the future. The non-Western intellectual project of futures must insist on exposing the political dimension of all knowledge relating to the future and cast the future not as an autonomous and inevitable domain but as a contested arena of conflictual practices—technological, global and scholarly—bound up with the perpetual expansion of the West. In theoretical terms, the project involves studying not what the future could or would be, but how new alternatives and options could be made to emerge and how alternative futures could be shaped according to the desires and visions of non-Western societies. In practical terms, the project has to focus on evolving a discourse of social involvement: in raising the future-consciousness of communities (including communities of intellectuals and academics), in articulating visions of societies and in involving citizens in efforts to shape their own futures.

In liberating the future, the non-West has a number of significant disadvantages. A great deal of emancipatory thought in recent years has concentrated on recovery of a discrete past. The non-West is coming to appreciate the creativity of its own traditions, and halting steps are being taking towards studying that creativity as a dynamic concept. However, that still leaves the problem of dislocation between past and present, let alone past and future. The past becomes, and for many Asian conceptual traditions has always been, a lacuna where ideals and aspirations reside. So a chasm is opened between appreciation of tradition and the imaginative capacity to think traditions forward. The imaginative leap is made even harder by the limitations of the language and techniques of futures studies and methodologies. A perception of a discrete past was intended to bolster the search for alternatives to Western dominance, to provide a means for continuity of values so that the non-West could move forward with its identity intact. However, it is a moot point whether the search for alternatives has not generated a public perception in the non-West of being the equivalent of opting out of the future altogether. Tradition is sold outside the technological wizardry of today and tomorrow. It provides perhaps a comforting answer to the dire needs of poverty, not a handle on power. Alternative technology became technology for the poor while the mainstream concern of 'real' development was getting a handle on modern Western technology. Critique of Western technology has so far failed to develop a concerted field of alternative technology products that offer a new nexus of possible lifestyles with market possibilities either for the non-West or the 'New Age' markets of the West.

The recovery of history has been a truncated endeavour in the non-West. Yet it does have the potential to upset the limited vision and self-satisfied composure of the future-forecasters. History holds a model for a different perspective on what is actually happening today, and from this can come new questions about how the future can be created. Corporate future-forecasters have leaped to embrace the so-called Pacific Century, the Pacific-Rim-centred view of the future. Much of Asia is preparing to dismemorialise the arrival of Vasco da Gama and incidentally recover the world his arrival disrupted. What stands behind this is an Indian-Ocean-centred vision of the world, its history and interconnections that opens a new perspective not merely on the distant past but also on present trends and hence the potential to engineer different, plural futures. Yet, how many alternative thinkers in the non-West are prepared to jettison the West as an integral part of their thinking and the centre of their future-consciousness? The trading world of Asia operated irrespective of the minor and distant European market. The trading world of Asia worked through plurality and interconnection through difference. Once more the trading world of Asia has the potential for sustained long-term growth by trading within itself. The fulcrum in the past was the Southeast Asian archipelago, today that again is a source of dynamic growth that can ripple and interconnect the old world of Asian interconnection. This is a rich topic for integration of past and future thought, but it has been analysed so meagrely that it has failed to generate much of an output to challenge the might of 'official' futures thinking. It is a timely conceptual basis from which to subvert the whole 'official' idea of how the future was made and will be shaped.

From this perspective, futures studies is not and cannot be a discipline in the conventional sense. Indeed, if futures studies was to become a fully fledged discipline, it would follow in the footsteps of ecology, cultural studies and feminist studies and be totally domesticated. Awareness of the future involves rescuing futures studies from any disciplinary constraints and from the clutches of tame professionals and academic bureaucrats. Futures studies must be openly incomplete, unpredictable and thus function as an intellectual movement rather than a closed discipline. It must work in opposition to the dominant politics and culture of our time, resist and critique science and technology (the most powerful agents of change and thought), globalisation (the most powerful process of homogenisation) and linear, deterministic projections (the official orthodoxy of the future) of the future itself. It must, in the words of Ashis Nandy, become 'a game of dissenting visions', 'an attempt to widen human choices, by reconceptualizing political, social and cultural ends; by identifying emerging or previously ignored social pathologies that have to be understood, contained or transcended; by linking up the fates of different polities and societies through envisioning their common fears and hopes'.[8]

For the genuine transformation of futures studies into a movement for resisting the status quo, its conceptual language has to change. Futures studies will remain alien to the non-West as long its basic concepts and categories are those of the dominant civilisation. This is why in non-Western societies, despite the best intentions of its practitioners, it often ends up subverting indigenous visions and futures. If the future is a state of awareness then that awareness can have genuine meaning only if it emanates from the indigenous depths of a culture. This means that there have to be a whole variety of futures studies, each using the conceptual tools of a particular culture and thus reflecting the intrinsic values and concerns of that culture. The plurality of futures has to be reflected in the plurality of futures studies.

Thus, intellectuals in non-Western societies will have to take the future seriously or become prisoners of someone else's future. They have to change the actualised future by changing the future-consciousness of their societies and by articulating the visions of their cultures in terms of their own notions and categories. It is probable that futures studies in different cultures will not be fully comprehensible across cultural borders; or, more particularly, that there will be incommensurability, in a Kuhnian sense, between indigenous notions of non-Western futures studies and Western futures studies. This incommensurability will arise from different norms and cognitive values, as well as different experiences, and it will be a product of the fact that many non-Western concepts and categories cannot really be rendered in English. Moreover, the incommensurability will itself be a source of resistance ensuring both a multiform of dissent and plurality of options for the future. Futures studies could thus become a genuinely high adventure generating a kaleidoscope of visions and fusing Other imaginations and moral concerns with political activism.

Problems often contain the seeds of their own solutions. Our awareness of the colonisation of the future arrives at a key moment in history, when the colonising civilisation has reached the end of its golden age. As ibn Khaldun would have said, the West, like all other civilisations, must now decline to rise again in some far distant future. The present phase of the cycle of rise and decline of civilisations favours the Asian civilisations. The next century belongs to Asia in general and India and China in particular. The centre of world trade has already moved to the Pacific Basin—the economic problems of Southeast Asia at the tail end of the 20th century notwithstanding. For the West, the growth of Asia could mean a return to a future of a thousand years ago. Both India and China, poised to become global civilisations, stand at the beginning of a cycle that could last a millennium; Western civilisation stands at the end of a cycle that is already a thousand years old.

But it is insufficient merely to accept the growth potential of Asia, enormous as that is and even though it is reconceptualised in the language of cycles. Alternative futures will genuinely emerge when Asia starts to think afresh by marginalising the West. That is the kind of equation Western dominance is working tirelessly to maintain as non-thought. Southeast Asia and the high-performing economies of East Asia were insulated from the recession of the 1980s by the potency of their growing economic interaction. This is a topic that does not figure largely in official futures thought. But it must be a starting-point for us, a willingness of Asia to think the unthinkable. Rather than be a victim of a totally colonised future, Asia needs to imagine that it can be a source of its own alternatives; that it can generate its own power base. The story so far is that those Asian countries that have most confidence in their long-term growth potential are also most in thrall to the power complex of the West; they are least able to see even their own power other than in Western terms and the language of Western futures studies. It seems that colonialism has predisposed the colonised to think only in colonial terms. That is the cycle that must be broken.

History and its cycles give us hope by serving as future-oriented memories. We can use our history and tradition to break the power of the present over our future. But first we have to equip ourselves to meet the formidable task. So, prepare yourself to rescue all our futures.

Notes

1. Warner Books, 1984.
2. Richard Slaughter, 'Looking for the Real "Megatrends"', *Futures*, 25(8), October 1993.
3. Oxford: Butterworth, 1992.
4. edited by Rowen Gibson, London: Nicholas Brealey, 1996.
5. Bethesda, MD: World Future Society, 1977.
6. edited by George Thomas Kurian and Graham T. T. Molitor, 2 vols., New York: Macmillan, 1996.
7. 'Colonising the Future: The "Other" Dimension of Future Studies', *Futures*, 25(3), April 1993.
8. 'Bearing Witness to the Future', *Futures*, 28(6–7), August–September 1996.

2
How the Future Is Cloned

KIRK W. JUNKER

In Don DeLillo's novel of American life during the Cold War, there is a scene in which J. Edgar Hoover, the Director of the American Federal Bureau of Investigation, is watching a baseball game. The year is 1951. During the game, one of Hoover's agents has informed him that the Soviets have tested a nuclear bomb. Shortly thereafter, still while Hoover watches the game, a piece of confetti—a page out of *Life* magazine—falls to him. The magazine page depicts Pieter Brueghel's painting 'The Triumph of Death'. Hoover studies the picture and reflects on the news, and it slowly becomes clear that this picture has set the tone of how he will frame the growing Cold War. 'There is the secret of the bomb and there are the secrets that the bomb inspires, things even the Director cannot guess—a man whose own sequestered heart holds every festering secret in the Western world—because these plots are only now evolving. This is what he knows, that the genius of the bomb is printed not only in its physics of particles and rays but in the occasion it creates for new secrets.'[1]

In the ancient study of rhetoric, 'occasion' was the term that students learned to answer the question of time and place—why speak here and now? In Chapter 1 of this book, Ziauddin Sardar has described the present occasion as a place and time for addressing the fact that the future has already been colonised and appropriated. Indeed this colonisation and appropriation has brought about the notion that there is only one future—an inevitably and scientistically Western future—for all of humanity. Ashis Nandy extends the occasion to require also that we address the fact that this one colonised and appropriated future disables and excludes dissent. In the following, I wish to explore this occasion, as identified by Sardar and Nandy, in an effort to see how we might enable multiple futures, including dissenting futures, by looking at images and stories. The first step will be to explore how we create futures to begin with. This creation is not a passive waiting for some inevitable material reality, but rather is an active, imaginative, narrative creation. The second step will be to see how we have come to create a present that seems to point to only one, inevitable future. Limiting ourselves to one future is a limitation that is enabled by living in only one present, which is in turn enabled by a Whig history that sees only one past—a past that, under the narrative constraints of 'progress' can only, inevitably lead to one present. In turn, as we give meaning to past events and people, by connecting them as data points on the line graph of a unidirectional, unilinear past-leading-to-the-present, we have no choice but to extrapolate the line (within error bars) to see one, inevitable future. Third, I will look at a specific future that we are in the process of creating through images and texts—the future of cloning. This leads to a concluding discussion of ways in which we can create multiple futures, including futures that dissent.

Throughout, I will use the images and narratives of cloning both as an example of how we have created a single future, and to show how we might create multiple futures. Cloning is a helpful example for several reasons. First, it is an example of the Western scientistic future, which, according to Sardar, is promising to be the only future, unless we take active steps to create other futures. Second, because cloning is controversial, people from all parts of culture (capable of creating all sorts of futures), have actively taken part in debating the issues and creating images and stories of what cloning means and will mean. Thus my tracing of the issues focuses upon popular culture, not scientific culture. Scientific culture has already predicted its cloning future—what others might be available? The stories and images of a culture, from popular as well as scientific sources, are part of the total meaning that we give to a phenomenon. Third, cloning is a helpful example because as cloning is a relatively new topic in public debate, its future possibilities have not yet been completely narrowed to one.

Creating Futures: Pre-diction, Pre-scription and Imag-ining

We have by now all recognised that there were multiple pasts, and that for a variety of reasons we choose to tell ourselves particular past stories as though they are the one and only past. Usually we do this to serve the needs of a particular present. And we can recognise that language can facilitate this selective telling of history. We even have a name for it—we call it 'Whig' history. The historiography of Michel Foucault, especially in his *The History of Sexuality*, has already persuasively demonstrated that the variety of pasts is evident, and our linear sense of history is itself just one story about one past, dominant though it may be. And there is plenty of literature that explores why we have chosen to tell ourselves this story as though it were the only story of the only past that leads inevitably to a one and only future. But how does this process of selective storytelling impact upon the future?

As with our narratives of the past, there are a variety of future stories to tell, and we will select to tell some and neglect to tell some. This selection process is not solely a function of the substance of the story, however; it can also be accomplished by the form of the telling. Here we will see that the form of the telling can change what appears to be a mere description into an imperative, and perhaps even normative, order.

Take the notion of prediction, for example. The ability to predict future events serves to create some power for the predictor, but, as Sardar has shown earlier in this book, it also serves to form that future, because the act of prediction exhibits a way of thinking that is limited to a certain cultural understanding. Moreover, the act of prediction is part of a cultural worldview that constricts the future to being only one future, by assuming that it is already 'out there' in some sense, waiting to be discovered. If multiple futures are possible, and indeed necessary, this notion of prediction therefore interferes with multiple futures. Steve Fuller puts it well when he writes that 'The love affair that Western thought has had with the idea of truth as something that is "discovered" or "revealed" finally comes to an end in the world of tomorrow.'[2] Therefore, knowing the future in a way that is predictable might not be desirable, if indeed it is even possible.

This constricting nature of prediction is not only due to its cultural ties, however; it is also due to its linguistic nature, in its construction and connotations. 'Pre-diction' literally means to speak before. Typically then, the sense of the word 'prediction' is to speak of an event before the event happens. If the speaking-before is sufficiently similar to the occurring-after, we say that the speaking-before is 'pre-diction'. But the judgement and interpretation that we exercise in comparing the occurring-after with the speaking-before is loose and accommodating. For many people, whose jobs or prestige or power may depend on being able to claim the ability to predict, from economists to weather forecasters to environmental planners, the accommodating looseness of matching the occurring-after to the speaking-before is happily used to make the occurring-after match the speaking-before; that is, to make events fit predictions. This is not to say that there is anything particularly wrong with this accommodating fitting—it is rather in the nature of the relationship of language to the world. But it is to say that owing to this inherent room for accommodation between language and the world, anyone who treats prediction as a scientifically precise endeavour reflects his or her needs, not the nature of prediction. Even within Western philosophy, David Hume made this abundantly clear long ago in his *Enquiry Concerning Human Understanding.*

Moreover, one need not even get the prediction 'right' to count. That is to say that the fact that we do predict, as a form of comportment towards the future, is itself a sufficient form to create futures—the substance of the prediction need not always be a match to the occurring-after event. Prediction thus becomes a large part of how we create both meaning and expectations. Ludwig Wittgenstein provides us with the example of someone saying 'leave this room' and a second person saying either 'I am leaving this room because you tell me to' or 'I am leaving the room, but not because you tell me to'.[3] An additional possibility, of course, would be for that person to not leave the room. The point is that once the first person has made a statement that we understand as being an order, no matter what the second person does, it will be interpreted as a response to the order. So too is it with prediction. Once a prediction has been made, we will interpret the occurring-after as being in some way either a satisfaction or a non-satisfaction of the prediction, but we cannot interpret it as wholly unrelated to the prediction. So by allowing ourselves to live in a present that features prediction as an important human task, we limit our futures to a future that is interpreted as a response to the prediction.

In addition, we add fuel to such singular future constructions by constructing the present in a way that suggests and perhaps requires that we predict the future. Our practices of economic forecasting, for example, create the reality of present currency and trade values. Unless we are prepared to alter radically our present practices to exclude prediction (which seems to be entirely unlikely), we must learn to work with prediction in a way that will still allow other, multiple futures.

Even more helpful than looking at 'prediction' in our examination of future-creation through language is looking at 'prescription'. Literally, 'pre-scription' is to write before. On strict word construction alone, 'prescription' should have the same sense about writing-before that 'prediction' has about 'speaking-before'. Typically, it does not, however. But if we treat it as though it does, we can arrive at yet a further insight about both. When we 'predict', we act as though the speaking-before in no way controls or affects the occurring-after—we are only guessing at what is independently 'out there'. Yet when we 'prescribe', we assume that we are talking precisely about trying to control the occurring-after with the very means of writing-before. Could it be that when we predict we are also controlling the occurring-after? Well, it is not as though simply saying 'it will rain tomorrow' will make it rain, or that by saying that 'tomorrow the British pound sterling will increase in value by ten pence against the Japanese yen' it will. But if we consider that, by predicting, we limit the types of discourse available for talking about the future, then indeed the act of predicting does control the future. And in precisely the mode in which Sardar warns, it limits the future to being of one type—we will say that the pound either did or did not increase ten pence in value at the end of tomorrow; and we will feel compelled to address the issue precisely owing to the fact that today we said it would. We might have understood tomorrow in an entirely different way if we had made no prediction. Prediction can, and does, have strong elements of command, of ordering and of prescription. I will discuss below how predictions about the future of cloning in fact can create the future of cloning. More importantly, we will see that the form of thinking about the future that we call 'prediction' also impacts on what cloning can mean, regardless of what the individual predictions are.

So we see that we can limit and enable different futures by language; that is, how we talk about them in the present. However, to return to the magazine page in the FBI Director's hands, how do images create the future? Need we be persuaded of the power of the image? Do we believe the cliché that a picture is worth a thousand words? If so, can that image determine a future in the way that we have seen that words can? It should come as no surprise that we believe the future will be, or will be created as, some manifestation of the past, particularly if we rely on visual imagery for that to occur. Before looking at images alone, I want first to address briefly the bridge from word to image in creating futures.

In associating images with language, Wittgenstein wrote, 'A picture held us captive. And we could not get outside it, for it lay in our language and language seemed to repeat it to us inexorably.'[4] Wittgenstein here suggests that the images are embedded in language and known through language. Appropriate to the present occasion, this questions whether we can even speak about a future without first having an 'idea' of what it might be—we create an image of what it might be. So we can accommodate the need to predict, or to know in advance, by expanding the possibilities to include imagining. Being able to speak about the future ('pre-dict') need not be an empirical account—it may mean creating images and projecting into them our beliefs and desires. If we consider the etymology of the English word 'idea', it becomes even more comfortable to speak of knowing the future through creating images. The word 'idea' comes to English, via Latin, from the Greek verb *idein*, meaning 'to see'. In about the 5th century, *idein* became equated with knowing as well as seeing. Similarly helpful is the word 'imagine'. We commonly connote 'imagine' as an abstract thought process, but its root is in 'image'.

Sociologist David Bloor provides an additional link for us between language and images that serves to extend Wittgenstein's point to the social realm. First, he notes that 'immersed as we are in society we cannot grasp it [society] as a whole in our reflective consciousness, except by using a simplified picture, an image, or what may be called an "ideology"'.[5] My present discussion of how we imagine the future fits into Bloor's notion of a theory of knowledge. Thus the relevance of his additional claim:

> The connection between social ideologies and theories of knowledge is no mystery at all but an entirely natural and commonplace consequence of the way we live and think. The social ideologies are so pervasive that they are an obvious explanation of why our concepts have the structures that they do. Indeed the tacit employment of these ideologies as metaphors would seem almost impossible to avoid. Our familiarity with their themes and styles means that the patterns of ideas that we have picked up from them will have an utterly taken-for-granted character.[6]

Bloor concludes by echoing the passage by Wittgenstein that opens this section: 'They [social ideologies as metaphors] will be unconsciously embedded in the very ideas with which we have to think.'[7]

Often when we advocate thinking about the future, or even creating a future, we do so with an idea already in mind of what that future, or those futures, should be, although the discussion may appear to be open ended. Must this be? Perhaps. If the image is embedded in language, and the language is pre-scriptive, then we can begin to see the normative aspect of creating the image. The image becomes a graphic prescription. It is very difficult to imagine (if we take imagining seriously) a future, or more radically, futures, without a particular future or futures in mind when we do so. To then suggest that what is in mind is in no way influenced by what we have already seen (the stuff from which we might compose an image) is difficult to believe. Here we approach the difficult thin line of having a future determined by the past, and, at worst, a linear projection of the past. And this is so even if we take imagining seriously. Thus the prescriptions of the future are not formed only in the present—what they could even mean in the present is constrained by the past.

To avoid unconscious or unintended linear-projected prescriptions into our images of the future, we need to break from the notion that there is just the one line of the past. If we re-examine the past as a variety of pasts, just as Sardar advocates a variety of futures, then no line, no matter how facile the graphist's error bars, can be drawn to include all the pasts. It is in this mind-set of a variety of pasts that we must begin to imagine a variety of futures, if we are to take imagining seriously, and more importantly, if we are to take our futures seriously. Only with the possibility of multiple futures can the one inevitable future make room for a dissenting future.

Limits of the Future Begin within Limits in the Present

'What does the future hold?', 'enter the future' and 'building a future' are such common ways of talking about the future(s) that we need to examine them as the recurrent stories that we tell ourselves about the future. For instance, in the phrase 'building a future' the notion that the future is created is not lost. But why do we limit 'creating' to 'building'? Building suggests a tangible image, and one that approaches the notion that a structure will be in place in which to put particular things.

Similarly, consider the phrase 'That's not what I had in mind'. How can we make such a statement? To think of an event or the description of a thing as being in mind, we speak as though the mind is a vessel that holds something, tangible or intangible. How real it must be that we can then compare it to something else, and say that the second is or is not the same as what is in our mind.

Sohail Inayatullah points out in this book that for some people the process of thinking about the future is enough to satisfy concern for futures. So long as we have 'raised consciousness' such that people 'think about the future', we should be satisfied. If we apply this abstract concern to the previous linguistic suggestion that we 'build' the future as a tangible vessel, then we can begin to see that being satisfied with simply 'thinking about the future' is tantamount to the creation of a future that is an empty vessel. In this way, like an empty vessel, the content of futures need not be discussed, so long as we are thinking about the future—that fulfils the expressed need to be 'forward thinking'.

So we can see that asking 'What does the future hold?' is insufficient for erecting multiple futures. We must interrogate this question and ask, 'Why do we speak of the future as "holding" something?' How does this metaphor of 'holding' both constrain and enable images of the future? This brings us to imagining the future not only as a vessel, but as an empty vessel. In Greenwich, millions of pounds have gone towards the building of a semi-permanent structure to celebrate the beginning of the next millennium in Christian time. This demonstrates that we are thinking about the future, but what will it contain? What images of the future will we create from which the anticipated millions of visitors will create the future? The answers were largely unknown when the structure was commissioned and when its building began. But how could this be? Does the unpredictability of the future prevent us from imagining one? What is it about the numerical change of a particular year that could cause us to build such a structure at so much cost without having first imagined what it would mean? Are we celebrating the fact that we survived the past or that we will see the future? We are already imagining that the building will have visitors and earn money. And we talk in this way. We are imagining that present accommodation and transport will be insufficient for the millions of visitors. So we will build more. That is then presumably what it means—because that is the meaning that we have given it: the celebration of the beginning of a new millennium will be marked by fabricating buildings of unknown use, reached by unknown means, visited by undetermined millions who will sleep in unknown places. The images, comical though they may currently seem, are none the less possible, and quite likely to happen.

Other empty vessels are all around us—the creation of web sites, for instance. Every person or institution with a sense of self-importance has a web site. What is on these web sites? The image of the future, as represented in a web site, is experienced when one clicks on what computer language calls 'icons' but often, contrary to the expectation of the user, no information appears, because none has been entered. University courses, industry training, faculty training and local computer stores can all provide the training for us all to create a web site, but who can provide us with something to put on it? If we apply Marshall McLuhan's adage that 'the medium is the message', we might discover then that the message is an empty vessel—an unattended web site.

What of universities? As the physical and mental locus of higher learning in the arts and sciences, universities are in their decline. In their place are distance learning, computer learning, the virtual university and universities of industry. Public and private universities are in large part already universities of industry that find it more important, owing to profit-margin educational economics, to obtain external funding from industry than to teach courses in the tools of living. Consequently, academics spend more time searching for funding than teaching courses. And when courses are taught, they labour under adolescent notions of relevance. Departments of business flourish and departments of literature grovel. As education becomes job training, the university looks more and more like an empty vessel that is used as a mould from which to fashion other empty vessels, but not from which to fill the vessel. 'Relevance' is treated as though alone it has meaning without asking 'relevant to what?' Consequently, any response to accusations of irrelevance comes in the form of providing more job training; that is, pressing out more empty vessels. Academics themselves have become so convinced of the one vision of relevance that they are most proud when they can boast a connection to industry, such as having obtained a grant, as though that is a measure of the value, quality and relevance of their work to the service of humanity. Physical scientists, social scientists and humanists are all guilty of this attitude. What remains of the university devotes its time to failing self-preservation and justification of its existence. Consequently, its substance has dried up and only the shell remains, like an empty vessel.

Before leaving this image of the vessel that holds the future, I am reminded that such a vessel might also be a vessel of resistance, and even possibly one of dissent. Ziauddin Sardar[8] has suggested that the balti is just such a vessel of resistance. He begins by explaining 'curry'. As post-war English tastes for foods from the Indian Subcontinent became more commonplace, Indian food came to replace chips as the lowest form of cheap food. Despite the variety of foods from the many corners of the Subcontinent, all became colloquially known as 'curry'. (This change could also be seen linguistically, when the word 'curry' changed from its adjectival modifier status to that of a subject-noun. Instead of eating 'chicken curry' or 'curried chicken', the sentence became 'I'd like a curry.') As a point of resistance to this generalising force on the variety of foods from the Subcontinent, restaurateurs obliged by giving the patrons all the same food and allowing it all to be called 'curry'. Escalation of taste discrimination followed. More sophisticated patrons wanted to be able to discern among the variety of foods, so they turned from the food itself to the vessel of preparation—first the tandoor, then the karahi, and then ultimately the balti. Just as with 'curry', Sardar points out that the balti can be a moment of resistance. It might also be a vessel of dissent, however, in that to the Indian, a balti is a bucket—a bucket that was used for all sorts of pail purposes, including the flushing of squat toilets. But it becomes a point of culinary distinction to the uninitiated coloniser; a place-holder to create a distinction in the preparation of food. Thus the vessel can be an image of resistance, and perhaps even one of dissent.

As tools for looking at how we construct futures, we have looked at prediction, prescription and images. Through both language and image we have seen that the future is actively created, not passively waiting to be discovered or revealed. Thinking of the future as waiting to be revealed is just one way of creating the future. I now turn to how we are actively creating the substance of a particular future, of how we are filling the vessel, with the future of cloning.

Cloned Futures

As recently, and as long ago, as 17 January 1994, the cover of *Time* magazine said 'Genetics: The Future is Now. New breakthroughs can cure diseases and save lives but how much should nature be engineered?' Since the future of now was already given (or colonised by) the assigned prescription of genetic engineering then, it should come as no surprise that our discourse has developed as though genetic engineering was the one inevitable future that would be increasingly revealed to us as the millennium rolls over.

Approximately one year after the news broke that Ian Wilmut and his colleagues at the Roslin Institute had successfully cloned a sheep (technically, it was nuclear transfer, or what Wilmut refers to as 'our technique', owing to his patent right ownership)[9] Wilmut reflected on the story in a lecture at Oxford on 3 February 1998, which he called 'Dolly: The Age of Biological Control'. Wilmut indicated that we are at the end of the era of physiological research, for which he gave the example of removing the pancreas to see how insulin was produced. He went on to say that the physiological era is beginning to be replaced with gene research instead. This provides insight into Wilmut's image of the future. Rather than rely solely on the story of making better human babies, Wilmut noted that cow's milk is already modified for help in its clotting factor. He then shifted from the present to the future, by extrapolating from the cow's milk to saying that it is 'likely' that pigs will be used as organ donors. Ultimately he did touch on the image of human babies. Lest anyone be naive enough to believe that people would not actually opt for something as seemingly debased as 'designer babies', we must consider the story told by the advertising image for the film 'Gattaca' when it was released in the United States. In the advert, one of the oh-so-cute babies in a nappy is pictured with the text 'Children made to order'. 'Only a tiny PG-13 and even smaller Sony Pictures logo suggested that they were not genuine. Calling an accompanying phone line, 1-888-BEST-DNA, prospective parents were promised the chance to "engineer your offspring". It was disconnected after the American Society for Reproductive Medicine appealed to Sony to change the ads.'[10]

In discussing what he called 'designer babies', Wilmut covered what he thought would be the five likely uses of genetic technology: to correct genetic disease, to reduce susceptibility to other diseases such as malaria, to improve IQ, to promote athleticism, and to improve aesthetic appearance, such as preventing baldness. Regarding the last three, Wilmut said 'I am personally frightened. I can't see how it would be practicable to regulate this.' Wilmut seemed to miss a point that John Glover demonstrated about those two uses that do not frighten him—the importance of the process of defining (in this case, 'disease'). If genetic technology is classified as acceptable for disease control, why can low IQ and lack of athleticism or hair not be called 'diseases'? When they are, they slip into the acceptable category. That is easier than trying to justify vanity and aesthetics. This relates to John Glover's 'solution', which we will see—to define eugenics unfavourably in order to exclude genetic technology. Wilmut's maxim is that we should be 'ambitious in research and cautious in application'. Despite the fact that these predictions frighten him, the narrative of what scientific discovery means relieves him of responsibility for these fearful images. During this same lecture, Wilmut announced that 'Some people think, quite erroneously, that I have some responsibility for what happens.'

During a lecture sponsored by Amnesty International at Oxford's Sheldonian Theatre in January 1998, the American philosopher Hilary Putnam said that in order to consider the possibilities for cloning technologies properly, we must create a 'moral image' for ourselves of what cloning will mean. This notion of imaging as an orientation to meaning is quite provoking, especially as it applies to providing meaning for possible futures. 'Imagining a future', taken literally, does not mean to think about the future in some undefined sense—it means to create an image of what the future will look like. And how will we do this? Can it be just a discrete picture of unknown meaning, put together like a police officer's composite sketch of the face of a possible criminal, pieced together from fragments of known faces and known things? No, a notion of a future is a notion that creates meaning. A picture of the future is a picture with meaning; an image of the future is an image with meaning. Just as DeLillo's J. Edgar Hoover used 'The Triumph of Death' as an image from which to create meaning in the Cold War, so too must we search for images from which to create meanings for the future and for futures. And insofar as we invest meaning in the picture that we imagine, we actively create a possible future, rather than passively wait for it to happen.

Putnam's observation was that the moral concern expressed by 'the public' over the cloning of Dolly the sheep was 'spontaneous'; that is, it did not seem to violate any existing moral principle or recognised right. This notion of 'spontaneous' is key here. The public of whom Putnam spoke did not require that there first be an identifiable moral principle, extending as a line from the past, next to which to lay cloning, in an effort to test for moral trespass. Instead, the public was able to determine an unacceptable meaning from cloning without such a mechanical step—it occurred 'spontaneously'. Critics might argue that this is undue media influence, or 'irrationality' at work. Yet it might also be that the present images created by cloning, if extrapolated, do not present any sort of desirable future image. This can be spontaneous and need not fit a past linear progression of a recognised moral precept.

Lawyer and medical ethicist John Harris disparaged the gut reaction that Putnam calls spontaneous because it leads to 'moral error'.[11] Putnam countered that spontaneous responses also lead to moral insight, since a gut reaction or spontaneous response must be in place before one can even 'be in moral business'. Here we can see the constituent nature of the moral image to creating a future that can handle the cloning of human beings. The spontaneous projection of a moral image, and reading back meaning from it, are a way of creating possible futures. This also serves to underline the importance of popular cultural images in the creation of possible futures.

The public and popular discourse focuses largely on cloning human beings, despite scientists' remonstrations that human cloning is not possible . . . yet. That same public's extrapolation of the cloned sheep to cloned humans was important, and should not be dismissed simply because we currently lack the technology. Because over and against this spontaneous concern is the image of technology that ultimately will enable human cloning. More important than the absence of present technology was the presence of an image that created a space for such technology to develop. Once that image was in place, stories abounded that gave meaning to that image of human clones. Instead of resisting this connection, we should ask how and why people would make such a connection—that is important. The importance of these stories and the futures of cloning that they enable and inhibit cannot be overestimated. In response to this foreseeable image of a future, Putnam invoked Kant's tenet that we are never to treat another person as a means. This moves the debate outside of images of infertile heterosexual married couples to images of such things as organ farms. Cloning someone in order to use the clone's bone marrow or kidneys would be cloning to treat them as a means.

Putnam finds that cloning human beings poses grave moral problems. Ian Wilmut himself admitted in his Oxford Lecture that 'there are questions of individual freedom' regarding cloning. We find the problems, says Putnam, by 'looking forward' first to learning how to clone and second to employing cloning not for infertile couples but for selective traits. Echoing Joseph Conrad's Colonel Kurtz's words in *Heart of Darkness*, Putnam asked why we view this scenario with 'horror'. Putnam fills the future vessel with the moral image of the infertile heterosexual married couple to guide the future. Objectionable though this simplified moral image may be it is at least not an empty vessel of 'thinking about the future' or 'consciousness-raising'. Lest professional philosophers should object to such use of moral images, Putnam was quick to point out that moral images are not a foundation from which moral principles should be derived. Justification of moral images can rather help complete a moral philosophy; that is, any moral philosophy without a moral image is incomplete, according to Putnam. Put another way, a moral philosophy without a moral image is an empty vessel.

What other images of cloning can fill the vessel? The consumerist image. But, insofar as the moral image is an ideal family, cloning does not fit into the moral image when it is a consumerist image—children become viewed as part of parents' lifestyles, like choosing furniture. Designer children, rather than just designer clothes, may result. Most importantly for present purposes, Putnam's moral image provides a willingness to accept diversity. This willingness to accept diversity is tantamount to creating multiple futures, or at least the possibility of multiple futures.

Still more popular images of cloning have been created that will guide the story of the future of cloning. In writing his thoughts on an interview with the filmmaker of 'Gattaca', Tom Charity falls immediately into the representation of the future through the image of the family when he tries to imagine ramifications of the story of cloning. 'The problem with cosmetic surgery is that it costs an arm and a leg. And that goes double for clinical genetics. But if we're unlikely to face the dilemmas envisaged in the above scenario, our kids may well have these options—if they can afford them. In a few decades, will our genetically modified grandchildren tune into Andrew Niccol's vintage 1998 sci-fi film, "Gattaca", and giggle at its naivety, or wonder at its prescience?'[12] Of particular interest here are several of Charity's choices of terms. First, is it not peculiar to have a 'vintage' in the present? Second, consider the use of 'envisage' to talk about how we 'face' the future. We face the future with our *visage*, our face. But this is not just a tautology—it is an image of our comportment toward the future, embedded in a word that also creates a sense of knowing, as with the Greek *idein*—to see and to know.

From 'Gattaca' we also see how language will control the meaning of what has already been discussed elsewhere—the material distinctions that could be made between those who are clones and those who are not. 'Gattaca' uses the terms 'valids' and 'in-valids' (which also gives us some insight into our own contemporary use of the latter term) to describe clones and non-clones; clones being valid and non-clones being either 'faith babies' or invalids. Employers in such a world are not interested in an invalid when they can employ a perfect individual—a clone. We do not need the film 'Gattaca' to know that insurance companies already practice that brand of discrimination. Dr David King, editor of *GenEthics News*, commented that 'People could be barred from certain jobs because of susceptibilities, rather than companies being forced to clean up their factories.'[13] In addition to speaking of the present as vintage, Charity makes a curious turn of phrase in describing the 'sight' of 'Gattaca'—'its futurism feels almost old-fashioned'. And as if being old-fashioned were not bad enough, he seems to feel that the ever-increasing speed of time demands that the present be shunned as well—'There's nothing new in "Gattaca"'s nostalgia for the present'.[14] Thus it seems rather clear that the film's future imaging focus pulls the interviewer–reviewer into treating the future as the present—literally.

The extension of the images of 'Gattaca' goes beyond just touching the interviewer–reviewer. The film's ability to create narratives that give meaning to the story of cloning was so immediate that one of the actors, while making the film, extended its language, and therefore its meanings, to the world around him. This actor, Jude Law, when talking of the making of the film, said the following: 'The Chicago Bulls were in the basketball play-offs with the Knicks, and we realised that the Bulls were valid, while the Knicks were in-valid: they were varying heights, and they would win games through their guts, sweat and blood, while the Bulls were these godly geniuses who only had to breathe to win.'[15] This is rather clear commentary on the ability of popular images to represent the stories that will give meaning to the future.

It is not as though a future can happen without images. The point is that we must interrogate the ones that happen. Even advocates of cloning recognise the role of imagery. In reviewing a speech made by John Glover, the director of the Centre for Medical Ethics at King's College, London, in which he attempted 'to rid modern cloning and genetics of the undeserved taint of Nazi eugenics', Justine Burley wrote that 'The Nazi ideology was one of racial purity and the Nazis used appalling imagery to give racism a biological justification, comparing Jews to vermin or to dirt and disease.'[16] Glover then indicated that the way to promote genetic engineering was to define it favourably and to distinguish it from eugenics by putting a spin on the possible images that could represent the practices: 'Glover believes that it is unfair and untrue to accuse parents who carry defective genes, and prefer to abort a foetus than pass those genes on, of practising eugenics. The definition of the term should be narrowed to exclude such cases.'[17] In picking up on the notion that the future is found in images, the first edition of BBC's monthly magazine version of 'Tomorrow's World', in commenting upon 'Gattaca', says that 'Writer–director Andrew Niccol's vision compellingly warns what a brave new world based on human genetics could be like.'[18] This short review inevitably leads to a promise of that same future: 'By the year 2000 scientists believe they will have mapped all 100 000 human genes.' We find two pages later that there is a reason to fear the vision, despite the reluctance of scientists to agree that this can spell nothing but cloning human beings: '"All animals are equal", suggests new research which shows that cow's eggs stuffed with DNA from sheep, pigs, rats and primates develop into embryos. The technique could mean that these animals, and even humans, could be cloned in cow's eggs.'[19] The headline article of that same magazine is announced on the cover as 'Welcome to the Future: 20 Visionaries Reveal the Headlines of Tomorrow'. Unsurprisingly, those headlines are all celebratory of science and technology, and never address human relationships. No explicit moral image is discussed, but one is implied. Dr Sandy Macara, Chairman of the Council of the British Medical Association, discussed the cloning of humans (it is not just the ill-informed non-scientist public who see it as inevitable) as inevitable, with no suggestion as to how we might do otherwise: 'The capacity to clone human beings could lead to prospective transplant recipients having themselves cloned so that their replica's organs could be harvested as spare parts. This would be attractive for someone with, for example, failing kidney function or leukaemia. The obvious ethical minefield should deter such horrendous developments, but history shows that when possibilities exist there will always be people ready and willing to exploit them for their own ends.'[20]

Despite protests from the biomedical establishment that human cloning is technically not possible, and therefore should not be entertained in public debate, on the other side of the Atlantic Ocean, the aptly named Dr Seed has provided an image of the future—human cloning. Strangely, for someone who claims no moral responsibility, Ian Wilmut made reference to Seed's announcement that he would be opening a clinic that offered cloning services to humans and stated that the suggestion made in Chicago (by Seed) was 'bizarre and appalling—this technique should not be applied to humans'. Despite Wilmut's protest against Seed, and his insistence that his nuclear transfer process does not facilitate cloning of humans at this time and is unlikely to be able to do so in the near future, and that as it is currently envisioned nuclear transfer could never reproduce adult human beings, Dr Seed has put a spin on public discussion that fits directly with what most people believe the meaning of cloning to be—cloning is about reproducing human beings and parts of human beings. Even Wilmut himself said that he could see 'genetic technology' used for human beings, and approved of it in disease control. Keeping the image and the story from Dr Seed's manipulations is no longer in Wilmut's control (even if the patent is) once he releases them.

In the end, Wilmut's protest is much like the modernist author who wants to claim his interpretation of his story is the only 'real' or 'correct' one. This is a story the reality of which cannot be controlled by the material limitations of stainless steel nuclear transfer devices. Need we be reminded that Mary Shelley's Frankenstein monster was just a story last century as well, but that its image is recalled again and again in this century by all sorts of audiences—from Nazi eugenics to Dolly the sheep. Frankenstein images are so common that they have all but lost their power to create an image, like Nietzsche's worn coin in 'On Truth and Lie in an Extra-Moral Sense'. When old stories like Frankenstein wear thin, we need to construct new stories to help with the creation and meaning of new futures, including those of the biological sciences.[21] But if we had been alive in the 19th century when Frankenstein was written, would we have thought it truly prescient of Mary Shelley to have described electricity as the force that started the monster's life? In describing the process of making Dolly, Dr David King reports that 'they took the nucleus from the udder cell and place it in contact with the "enucleated" egg. Passing an electric current through the liquid made the membranes surrounding the nucleus and egg fuse, so that the nucleus entered the egg. The electric current also stimulated the egg to start dividing'.[22]

Another image and enveloping narrative on genetic research and technology from within the medical establishment is presented in the opening paragraph of *A Quest for the Code of Life: Genome Analysis at the Wellcome Trust Genome Campus*: 'Medicine is in the midst of a revolution. Modern genetic research is radically changing our understanding of the factors that make us prone to disease, as well as those that determine, at least in part, the individuals we are destined to become. The key to understanding the very basis of life lies at the heart of almost every cell in our body; it is literally in our genes. Cataloguing each of our estimated 100 000 genes—the "recipe book" for a human—is the ambitious goal of the Human Genome Project. Described by some as biology's equivalent of sending a man to the moon, the Human Genome Project will provide the information that will transform medical care in the twenty-first century.'[23]

Interestingly, Wilmut himself raises the question whether one could 'have a normal healthy relationship with a child who is a copy'. Speaking from his experience with his own two children—one of whom is adopted—he answered the question in the negative for this reason, and because at this juncture, one year after the cloning of Dolly, Wilmut claimed that no superior reasons had been given for cloning humans.

Thus it seems to be rather clear that despite the absence of present technology capable of cloning human beings (for any purpose), the presence of Dolly, among other images, has created sufficient opportunity for image and word about the future to justify considering the moral issues of cloning. Our task then becomes one not of denying this possible future—the words and images will midwife it into being—but of making it only one of a multiplicity of possible futures. To do this, more than the mode of scientific material realism must be given space to colour the image and must be given voice to speak the word.

A Multiplicity of Futures: Pictures at an Exhibition

In his reflection upon technology 'On the Question Concerning Technology', Martin Heidegger repeats the words of the poet Friedrich Hölderlin: 'But where danger is, grows the saving power also.'[24] So from the danger of a linear past that leads through the present to the one, inevitable, predictable future, grows the variety of pasts, the chosen present and the multiplicity of futures—including one of dissent. How do we represent multiple possible futures? Typically, we do so by juxtaposing 'reality' with something unlabelled—left therefore to be understood only as something other than reality. So, for example, when *Time Out* magazine reviewed 'Gattaca', separate from the review, but on the same spread, was a corresponding interview with Dr David King, editor of *GenEthics* and putative spokesperson for reality. However, there is more than one future possible reality—if we are creating, rather than discovering, futures that is.

Likewise, at an exhibition called 'Multiplicity', young artists were asked to provide a graphic interpretation of what they thought cloning meant. The pictures alone provided images, and the artist's commentaries on their own work were also present, from which a visitor could read the artist's own verbal meanings. These paintings are images that reflect the reality of present stories that we tell ourselves about cloning, and therefore help to shape its future. And to foster that relationship, each painting included a brief text by the artist. 'Kopy Kat', one of the most discussed pieces in the exhibition, depicted three cans of cat food each containing 'one Kat'. The artist, Joanne Welsh, asks, 'Why assume that genetic engineering will lead to the calamitous? Just as likely, the mundane.' In a similar mood, artist Mark Farhall wrote, 'The inevitable future of cloning. Frivolous.' Here then were variations on the future with cloning that did not express some of the spontaneous fear to which Hilary Putnam alluded, but instead envisioned grey sameness and frivolity.

Another future that the artists explored was the industrial future of cloning. Tim Ellis entitled his painting 'Copy. Right?', not missing the fact that Ian Wilmut's team has already patented its nuclear transfer 'technique'. Taking his image of the future from the industrial unilinear story of the past, Akash Bhatt created an Escher-like image called 'Production Line'. He commented that 'Cloning human beings to the point where individual identity is lost is like that of a production line where the next one is like the last.' The fear that the 'Gattaca' film evoked was present in a work by Sophie Dauvois entitled 'would you like to have a clone?' This multimedia work consisted of a distorted mirror with a voice recording saying 'I am the original, you are the clone.' The papier mâché frame repeated the pattern of the four bases of DNA (GATC) in various combinations. The 'Multiplicity' exhibition concluded with social science evaluation forms (multiple-choice of course) and a comments book for visitors, many of whom decided to draw their thoughts as pictures as well as words.

At the beginning, or end, of this circular arrangement of twenty-four pieces, the visitor came to 'the reality of cloning'. In a case were tools of the type used in nuclear transfer, a book on cloning, and a shock of Dolly's fleece. But these were not the only items in the case. At the top of the identification card in the case, and signified as 'Exhibit A', were 'Guaranteed Dolly Droppings'. The sheep faeces were placed in and beside (as if Dolly had, with the carefree aim of one accustomed to dropping droppings in the field) a powder-blue Tommee Tippee baby potty chair. Here then we see the symbol of the benefactor of cloning—our children's children, male and Western—in relation to a symbol of the 'reality' of cloning itself. The common act of children urinating and defecating, with which parents are all too familiar, symbolised in the little boy's plastic potty chair, is mixed in symbolic unity with the common act of a sheep, a real sheep, whose reality and commonality with naturally inseminated sheep is emphasised and symbolised by its real faeces, to provide an image for the future. In discussing the 'Multiplicity' exhibit, London's *Time Out* reviewer Julia Thrift concluded with 'The fetishisation of sheep shit seems to sum up our confusion about Dolly and what she represents. I wonder what future generations will make of it?'[25]

The narrative to place the paintings in a context is provided by Wellcome: 'Using dependent language, the future can be created: One practical application of this science [genetic modification] is the development of transgenic animals (containing genetic material transferred from one species to another): for example, sheep that could lactate human proteins for use in the treatment of cystic fibrosis and haemophilia, and pigs that could provide organs for transplantation to humans.'

Conclusions

The dynamic relationship between the fantastical picture 'The Triumph of Death' and the event of the nuclear bomb explosion provides an excellent example of how images can give meaning to, and guide futures. An advantage of multiple futures is that we are not wedded so tightly to one way of thinking that when it fails us, we must stick with it anyway or face drastic, chaotic alternatives. Think of having a family to fall back on when a job or marriage fails, or a job to occupy one's time when a marriage is failing, or a spouse to whom one can turn when family relatives or one's work are intolerable. Creating futures will include the Western future not because it is inevitable (it is not) but because foreclosing it will be committing the same error of single future.

What is the reality of cloning? Is it the tools of nuclear transplantation? Is it the determinative material of genes—DNA? To proponents of genetic engineering and cloning, these materials are the 'reality' of the science that points to the future of cloning. But what about what people make of this scientific material reality? Is it not also one of the realities of the future as suggested by cloning? The stories we tell ourselves about the past, present and future are as much the reality of human experience as the material objects that play roles in the stories. To ignore them as realities is to suggest only one reality—a linear scientistic reality, steeped in the notion of progress, from which we predict the future by trajectory from the past (so long as we keep the datum point known as the present within acceptable error bars).

Notes

1. Don DeLillo, *Underworld*, New York: Simon & Schuster, 1997.
2. Steve Fuller, *Philosophy, Rhetoric and the End of Knowledge: The Coming of Science and Technology Studies*, Madison: University of Wisconsin Press, 1993.
3. Ludwig Wittgenstein, *Philosophical Investigations*, 3rd edn., translated by G. E. M. Anscombe, London: Basil Blackwell & Mott, 1958, p. 487.
4. Ibid. 115.
5. David Bloor, *Knowledge and Social Imagery*, 2nd edn., London: University of Chicago Press, 1991, p. 53.
6. Ibid. 76.
7. Ibid.
8. Ziauddin Sardar and Borin Van Loon, *Cultural Studies for Beginners*, London: Icon Books, 1997.
9. Ian Wilmut *et al.*, 'Viable Offspring Derived from Fetal and Adult Mammalian Cells', *Nature*, 385(6619): 810–13, 27 February 1997.
10. David King, 'Gene Genius', *Time Out*, 1438: 13–15, 11–18 March 1998.
11. John Harris, *Times Higher Education Supplement*, 23 January 1998.
12. Tom Charity, 'Cell Shock' (interview of Andrew Niccol), *Time Out*, 1438: 12, 11–18 March 1998.
13. King, 'Gene Genius', p. 13.
14. Ibid.
15. Ibid. 14.
16. Justine Burley, 'Hitler's Spectre Haunts Science', *Times Higher Education Supplement*, 6 March 1998: 22.
17. Ibid.
18. '2020 Vision', *Tomorrow's World*, April 1998: 9.
19. Ibid. 11.
20. Ibid. 26.
21. Jon Turney, *Frankenstein's Footsteps: Science, Genetics and Popular Culture*, New Haven, CT: Yale University Press, 1998.
22. King, 'Gene Genius', p. 15.
23. Liz Fletcher and Roy Porter, *A Quest for the Code of Life: Genome Analysis at the Wellcome Trust Genome Campus*, London: The Wellcome Trust, 1997.
24. Martin Heidegger, 'On the Question Concerning Technology', in *Basic Writings*, edited by David Farrel Krell, London: Harper & Row, 1977.
25. Julia Thrift, review of 'Multiplicity' exhibit, *Time Out*, 1438: 45, 11–18 March 1998.

3
Rethinking Futures Studies

ELEONORA B. MASINI

Futures studies has a history that can be traced to the end of the Second World War, although of course thinking about the future has been a central activity of men and women since the beginning of civilisation. People become human the moment they think about the future, the moment they try to plan for the future. The future is a symbol, as John McHale used to say,[1] through which we order the present and give meaning to the past. But attitudes towards the future have certainly changed in history at different times, in terms both of how people look at the future and of how contemporary values affect future perspectives. It is interesting to take a brief glimpse at some of these changes.

In the *Republic* of Plato, the vision of a future society is one based on justice. Saint Augustine's *City of God* is a society based on love and pitted against the 'City of Man' based on pride. Augustine contends that his perfect society can be made real through structural change in the City of Man. We can look also at the *New Atlantis* of Francis Bacon, a society based on human greatness, and at Thomas More's *Utopia*, in which communal ownership of goods is central and where the individual is subordinate to the community. And we can proceed from these examples to consider the social ideals put forward by Comte and Marx to solve the pressing social problems of their eras; and examine the ideas and experiments of scientific utopians and social reformers since the 19th century.

All these different visions of a desirable future are in some way embedded in the social structures from which they emerge, and are linked to the needs and the hopes of the people living at the time. We, in the present time, need to look at the future in ways that go beyond the creation of beautifully conceived but ultimately illusive utopias. Our future must be not only foreseen and dreamt of, but also chosen and built.

Let us now trace, in very general terms, the historical development of futures studies, from the Second World War on, and then try to understand its philosophical basis.[2] Since the end of the Second World War, mankind has sought to tackle the ever quicker and more interrelated transformations taking place and to identify the future consequences of present actions in order to avoid being overwhelmed or taken unaware by events. The endeavour to anticipate events through scientific analysis of trends and indicators of change—the technological forecasting component of modern futures studies—developed first in the United States during and just after the Second World War. Soon thereafter, Bertrand de Jouvenel and others in Europe began addressing the philosophical and sociological dimensions of futures studies, stressing the importance of forecasting alternative possibilities and considering in detail the long-term consequences of current policies and actions.

With time, it became apparent that foresight was important not only in order to know where one was going and how, but also to choose where one *wanted* to go. That is, by identifying futures that were possible and perhaps the most probable, futurists could progress to think about desirable futures. But 'desirable' on what basis? On the basis of an individual's choices, or on the basis of choices by groups, cultures or ideologies? Hence, more recently, futures studies has seemed to become increasingly linked to philosophical choices, to choices of principle and to choices of how one is to regard reality, man and society. Unfortunately, this development is not strong enough, and certainly not the declared goal of all futurists at this time. But I want to stress the importance of basic philosophical choices in futures studies.

The Evolution of Futures

In the first period following the Second World War, the French futurists studied and investigated the scientific and political aspects of futures studies. In the latter half of the 1950s, in fact, Gaston Berger founded a centre for 'prospective studies'. The term 'prospective', as it began to be used in Europe, referred to making decisions based not only on immediate needs but also on long-term consequences. After Berger's premature death, economist Pierre Masse continued the 'prospective' effort. As general commissioner of the French national development plan, Masse was responsible for giving prominence to the prospective mode of thinking —a prominence that led to the adoption of the first French national plan for 1985. In the same period, Bertrand de Jouvenel gave the futures movement an extremely important boost with his studies on power, methods of governing and political choices. The central feature of his thought is the overall dimension of time—past, present and future—which alone can enable man to function in the political arena. De Jouvenel continued to work on this research theme; and to this end he founded the Association Internationale de Futuribles, which acts as a clearing house for research on the future.

Other European counties have also made important contributions to the field of futures studies in terms of philosophical bases. In the Netherlands, sociologist Fred Polak theorised the birth of futures research from an epistemological point of view in his books *The Image of the Future* (1961) and *Prognostics* (1971). Between 1966 and 1972, the political establishment in the Netherlands demonstrated some interest in futures studies, even establishing a policy sciences unit in 1974. The work of Jan Tinbergen's group is also extremely important, particularly the report it prepared for the Club of Rome on RIO, the New International Order. In Great Britain, futures research was pioneered by the Science Policy Research Unit (SPRU) of Sussex University. In addition to presenting criticism of global models, the unit attempted to create a theory of futures research with contributions from an interdisciplinary team. The Scandinavian countries too developed futures studies and offered the findings of these studies to their governments, and to research institutes and organisations, in an attempt to arouse more interest in the future. In Laxenburg, Austria, the International Institute for Applied Systems Analysis (IIASA) attained considerable prominence throughout Europe. This institute was created with the financial and scientific co-operation of various national academies, particularly the United States Academy of Arts and Sciences and the Academy of Sciences of the USSR. The institute's work focused on the application of systems analysis to different fields, most importantly in the energy sector.

Many international groups have also been involved in futures studies. In addition to the Club of Rome and the International Futuribles Association of France, two other futures groups of international interest emerged in Europe—Mankind 2000 and the World Futures Studies Federation (WFSF). Both groups were given strong impetus at the first International Futures Research Conference, held in Oslo in 1967, which was organised by the Peace Research Institute. The aim of Mankind 2000 (founded in 1966 in Holland, and later moved to Brussels) is to promote all aspects of human development in individuals, societies and in emerging world communities. The central purpose of the WFSF is to serve as a forum for the exchange of information through publications, conferences and meetings. The WFSF first focused its members' activities on the study and analysis of human needs in terms of future societies, and gradually became more interested in communication and cultural identity. Currently, it is working on evolving a knowledge base for futures studies, establishing norms and criteria for professional futures work and developing visions of alternative futures societies.

The former 'socialist' countries of Eastern Europe merit some special remarks for their efforts in futures studies. The term used in these countries, before the fall of the Berlin wall, was 'prognostics', which considered futures studies as the crucial process preceding the formation of a plan. It is worth expanding on this definition.

The foundations of socioeconomic planning in the former Eastern bloc states were laid according to the regularities of the modern scientific–technological processes. Here the links with the positivist tradition and with Lenin's thought are evident: Lenin made the principles of scientific communism the basis for planning. According to his school of thought, forecasting is the stage prior to planning. Forecasting and planning differ in levels of objectivity and complexity but are necessarily tied. On the basis of a dialectical-materialist outlook, the future (as distinct from the past and the present) is in principle stochastic and not simply a projection of the past; indeed, the future contains within itself a large measure of creativity. The socialist countries, therefore, concentrated on the analysis of the scientific and technological process and on the consequences thereof as elements of social progress. Since the break-up of the Soviet empire, this approach to the future has been largely abandoned; but its historical significance cannot be overstated.

The first future studies in the United States were invariably conducted for government agencies and business corporations. Only later were such studies carried out by universities and private research centres such as the Center for Integrative Studies directed by the late John McHale at the University of Houston and later at SUNY/Buffalo by Magda McHale. Systems future studies are currently being conducted by Harold Linstone in Portland, Oregon, and by James Dator in Hawaii, who has involved local citizens in efforts to design their own futures. The varieties of futurists in the United States and Canada, from both historical and contemporary perspectives, can be tentatively divided into the following categories:

- **Technologically oriented.** These pursue future studies based on the use of technological processes. Representatives of this group include the late Herman Kahn and Olaf Helmer, and Theodore Gordon.
- **Sociologically oriented.** Alvin Toffler, James Dator, Daniel Bell, and the late John McHale in the United States, and Kimon Valaskakis in Canada seem to fall conveniently within this category.
- **Globalistically oriented.** This group includes those futurists who have worked on the Club of Rome's various projects (such as Dennis and Donella Meadows and Mihaijo Mesarovic) and those associated with the World Order Group in New York, directed by Saul Mendlovitz.

Mention should also be made of futures studies in the developing counties, which have shown an increasing interest in the field. Within the Francophone region of Africa, Morocco was a leader in futures activities; in the English-speaking zone, Egypt, Kenya and Tanzania have well-developed futures communities. Both India and Sri Lanka have highly respected futures groups, and of course Japan follows the developed countries' pattern and interests. Latin America has recently been expanding its activities in futures studies, particularly in Mexico, Venezuela and Brazil.

Philosophical and Ethical Foundations

Following this very brief description of the past and present of future studies, it is important to analyse the general philosophical orientation that various groups involved in futures research have followed; even though these assumptions may not always be conscious.

In the French school, the primary focus of interest has always been the consideration of 'alternative futures'. There is not one future; there are many futures. Which of these futures emerges depends on the choices made by human beings. As Michel

Godet describes it, 'the prospective approach ... reflects awareness of a future that is both deterministic and free, both passively suffered and actively willed'.[3] In the Netherlands, Fred Polak's approach is very interesting—namely that the future is based on images of the future that are related to historical situations, belief and desires. To understand the future, Polak argues, we must examine its premisses in people's minds.[4] Other futurists in the Netherlands are oriented more towards searching for trends that would lead to policy choices. This is also the orientation of many Scandinavian future studies: towards praxis. In Great Britain the directions are varied but are generally oriented towards perceiving the world as a whole, taking into account relations among individuals, between man and society, and among societies. British futurists are particularly interested in the impact of modern technology on societies. Also of particular interest is the work of the IIASA in Austria as it attempts to relate specific topics to the global system as a whole, as in the case of energy.[5] But these differences apart, it is important to notice that the approaches of most futures studies, in the recent past and even today, are related to Western philosophical concepts—particularly those of John Locke, G. W. Leibnitz, Friedrich Hegel and Immanuel Kant. But I consider it important to trace not only the philosophical foundation of futures studies, but also their ethical foundations.

Futures studies involves the possibility of looking into the future at various levels in order to better understand the changing interrelations between man, society and the environment. The three levels of looking into the future reflect three different philosophical approaches to the future. The first approach emerges from the need to face rapid change and to know where the world is going. It consists of the data of the past and present, which point the way towards what is possible and help identify what is probable among what is possible. It is based on the proposition that 'something *is* changing'. This 'prognosis' approach, which relies on extrapolation and makes heavy use of social and economic indicators, was widely used from the end of the Second World War to the 1960s; and it has now re-emerged as the dominant form of American futures studies. This approach is most closely related to the philosophy of John Locke, based on empirical data.

The second approach is linked to utopias—desirable societies. In terms of futures studies, this means that the future is built on the basis of something we wish to happen. While in the 'prognosis' approach the 'possibilities' and the 'probables' are sought, this second approach seeks the 'desirables'. I call this the 'visions' approach because it aspires to transform the present by a *vision* of the future. The philosophical basis of this approach comes from Leibnitz, and is based on the belief that 'something *must be* changed'. I believe this approach to be very important. Futurists must think not only of 'possibles' and 'probables' but also of 'desirables'. We have to examine the forces that draw us on; otherwise we will think only of what has been done before, and change will not occur.

The third approach to futures studies is a synthesis of the first and second. It is on the level on which people think about the future in terms of *projects*. This means they seek to undertake projects that will change reality according to specific indications directed by utopias, by social ideals, by models and by visions, while at the same time taking into account empirical data on trends in the past and conditions in the present. This third approach is based on both a knowledge of 'possibles' and 'probables' and a vision of 'desirables'—on models and on ideals. Here we see the emergence of choice—the interest of the observer. It is based on the belief that 'something *can be* changed'.

From among the 'possible' and the 'probable' emerges the 'desirable' (in Lantian terms 'the ideal', and in Hegelian terms 'the infinite'), inserting itself into reality and creating a synthesis, or rather, a synergy. The project that emerges is based on assessments of the possible and the probable and the choice of the desirable. I call this approach 'project-building'. I believe that it may also relate to non-Western philosophies, but I have not as yet pursued research into this very important area.

Futures Thinking as Vision

As I have said, the visions approach focuses on desirables and emphasises values; although values are always present in every approach to future studies. Futurists using this visions approach often rely on the existing visions of writers, poets, political scientists, philosophers or even policy-makers. But they could, equally profitably, develop their own visions.

Visions spring from the capacity to recognise the seeds of change that lie in the past and the present; moreover, visions make it possible to create a future that is different from the present although its seeds are in the present. In a sense, visions capture the changes that are already latent in the present and posit these as the future reality. Visions are born from the capacity to listen, to search, to be attentive to that that already exists, but that is not yet obvious and may develop later. Visions are linked to people who carry the seeds of change, and are not mere abstraction. The ability to nurture the seeds of change and develop visions is even more important than the capacity for future analysis. For example, when an important element of change (such as a different lifestyle) develops, the people involved may be simply carried along, unable to do anything about it or sometimes not even aware of it. In this sense people are not part of the process and do not choose to change but simply accept it. This process continues and reinforces itself because the fewer the number of people choosing the changes, the less consciously directed change becomes part of the change process.

Instead, it is important to perceive what I have called 'the seeds of change', those aspects of society that are in the process of developing and that require new modes of understanding that go beyond the rational and work at the levels of intuition and emotion. Often the unconscious mind offers understanding that is richer even if less explicit than those understandings provided by logical analysis alone.

The ability to recognise the seeds of change is stunted in people who accept change in a passive way or who are simply interested in keeping alive and expanding the social system of which they are part.[6] Such people are locked into their social character.[7] This term is used by Erich Fromm to mean the nucleus of the character structure that is shared by most members of the same society. In fact, such members of society will want to act as they are expected to act in the social system to maintain it. The social character as such becomes an assumed identity of the individual, which is necessary for the perpetuation of any social system.

Who are the people capable of capturing such latent changes, the people futurists must listen to in order to become vision-oriented futurists? The best listeners, capturers of seeds of change, are those who do not fit the existing social character in its totality. The capturers, the listeners, are those who somehow are outside the logic of that specific system. Some would account for these people with 'maladaptation theory'; but I wish to go beyond such a simplistic explanation. In fact, if we consider the artists, the writers, the poets or the politically persecuted in oppressive regimes as the 'outsiders' then we have to reject the maladaptation explanation altogether.

The listeners—the capturers of the seeds of change—are those who build visions that are different from the present but are not pure utopias since they are part of the process of history. These perceptive individuals are mainly those who do not perceive only with their rational capabilities but also with their intuitive, imaginative capacities. This is the case of artists, poets, philosophers and others who exist outside the existing social system, which mainly tends to conserve and perpetuate itself.

There are other groups that are not part of the system in the sense that they do not help perpetuate it, and these groups include, I believe, women and children. I shall talk about children first. They have to accept the social system of which they are a part but they are also silent witnesses to change. They are, if listened to and stimulated, capable of providing visions for a different society from the one in which they live. Of course, if they are simply talked to in terms of the social system of which they are a part, with its vocabulary and its frame of reference, they also seem to conform. But it has been shown by research in Italy that when the methods of listening to children are identified—even if in a tentative way—new visions emerge.[8] Women too, I believe, are listeners to the seeds of change in the process of history. Because women have not competed with men until recent times, particularly in the Western world, they have not become completely captured by the social system and hence have not completely absorbed the prevailing social character. In fact, women have acquired and kept alive the capacity to build visions. This is shown by the opposition that women have voiced to the predominant emphasis in the West on economic priorities over the quality of life for human beings. This point has been expressed in global terms through the visions of alternative futures offered by Hazel Henderson[9] and of a world of peace by Elise Boulding.[10] Women have lived their 'alternatives' to the mainstream of life in their private lives, and the indications now are that they will continue living them in public. The world of today (and probably of tomorrow) shows signs of stress and even of possible catastrophe. Perhaps women's visions, based on perceptions that are not purely rational, can create an alternative that may ensure the survival of humankind.

It is crucial that the visions approach to futures studies stop being an evasion and start to be recognised as a force by those who are bearers of the vision. If the capturers of seeds of change are aware of the importance of their visions to the process of history, they will also acquire status and believe in themselves. In this sense their 'visions' will no longer be 'utopias' that, though linked in their premisses to the historical moment when they are devised, are in fact evasions of the present —societies that parallel reality and have no force for change. This has been the case for many utopias in history. The importance of 'visions' must be understood by futurists who have mainly relied on analysis and extrapolation, and on knowledge of the past and the present. Visions of truly different futures have been treated by most futurists as of secondary importance—as not very serious products of fiction writers or philosophers. As a mater of act, futures studies has not even realised that extrapolations are tied also to visions, though in an obvious manner—as has been described in the analysis of so many global models. Futures studies has to re-evaluate visions and search for the listeners to seeds of change wherever they are—in art, in alternative political movements, among women and children. This will make their forecasts more than mere extrapolations—predictions more linked to unexpressed needs. Some writers speak of the peripheries in respect to the centres of power while others speak of the minorities and the mainstream. However we wish to label them, these groups are now very large: for they include most people of the non-West and the women of all the world.

Project-Building: The Responsibility of the Futurist

I have stressed the importance of visions in futures thinking and also the fact that such thinking must be rooted in the changes that are emerging as seeds in the process of history. But it is also important that such visions become, together with the analysis of past, present and future trends, *projects* of action for the future. It is not enough to identify those trends that may develop in the future, or to have knowledge on which to base analysis. Nor is it enough to have a vision even if it is rooted in the historical process. What is important is to build with the aid of trends a project that will then bring about the vision. It is at this point that the future can become what we wish it to become, or at least this can happen if unknown elements do not emerge. This is the moment when the element of force is important—the element of will that follows choice. It is the moment when strategies and even tactics must emerge, and it is this point that is based on all elements: desires, possibilities, probabilities and will.

In such terms, the concept of 'project' has yet to be explored. For the most part there are either short-term projects about what way to develop or there are grandiose proposals for the future such as those of many ideologists. What is needed, though, are projects with a basis in past and present knowledge, and futurists who are aware of the short term but capable of seeing things in the process of change and who are inspired by a long-term vision.

The frame of reference—the value system in which the futurist operates—needs to be clearly defined, for it is the futurist's responsibility both to exercise rational judgement in influencing decisions and at the same time to be creatively self-expressive. The expression of oneself is clearly related to one's values, while the rational influencing of decisions must include sensitivity to the ideas and values of others.

If we analyse the three approaches to future studies described above, we can see that values and ethical choice are more or less evident. The future is the only temporal area over which people have power: the past and the present are always beyond control. These areas may offer knowledge, but no more; they are the 'facts' that can be utilised for futures that in their turn are built by human will—meaning that the future emerges from choice among the various models of reality that humans wish to build. The future emerges from the chosen model of a specific historical moment, whatever the approach, although this emphasis is more apparent in the visions and projects approaches. How can we perceive these various historical–philosophical approaches?

Is it enough to know data and construct a model, giving importance to experience, trends and extrapolations? Conversely, is it ethically defensible to build a model for a utopia and *not* seek to transform it into action? This is a topic that merits more than philosophical discussion; for at this point the social and the ethical responsibilities of the futurist merge.[11] The futurist is part of the world and of the dynamics that he or she is trying to describe in future terms. The futurist cannot build himself or herself a comfortable ivory tower like some theorists in the so-called 'pure' sciences. The futurist, in fact, 'reflects' his or her own culture and, at various and different levels, his or her disciplinary matrix. On the one hand, he or she believes that he or she interprets and reflects his or her culture completely.[12] But in reality, this reflection is only partial, because together with the cultural elements, the futurists also reflect their social character and perhaps their individual temperament. On the other hand, sometimes the futurists reflect their culture without knowing it; they think they are speaking for the world and forget that it is their cultural biases, their disciplinary education and their social character that is being expressed—and that these are only partial aspects of the world. Futurists, more than any other scientists, need to acknowledge the existence and the value of cultures, attitudes and objectives that are different from their own. Hence the importance of what Mihai Botez calls the implicit hypothesis of the parts that are 'excluded' more or less willingly.[13] These considerations indicate that the futurist cannot be considered a pure technician but is the bearer of a set of values that are both declared and undeclared—a bearer of implicit and explicit hypotheses. Bertrand de Jouvenel has stated that observation is related to knowledge and to the interest of the observer.[14] This is especially true for the futurists, who, because their field contains much more than they can observe and because it deals with what has not yet occurred, must put questions to themselves that range beyond their own interests and try to reduce their area of 'non-observation', in which the unforeseen —the 'non-forecast'—may prove fatal.

This aspect of the work of the futurist must be understood if we wish to re-conceptualise futures thinking to include developing and searching for visions and building projects. Futures thinking must be linked to social responsibility and to ethical values that are clearly expressed and defined. This does not imply setting rigid definitions that constrain outlook, but merely that the futurist make his or her values open and understandable, and hence recognised, even if not approved.

In fact, projects for and of the future must be many, reflecting different values and appreciated for their diversity. Often what are actually projects for the future are presented as if they were only extrapolations: the future is going to be such and such. On the other hand, sometimes what are presented as visions are actually projects for building a future that is different from or opposed to the present. But whatever their disguise, projects are ethical statements, acts of will, that futurists wish to be translated into reality. At their ethical best, such projects are not simply consequences of the past, nor schemes to perpetuate the present or to fulfil a private desire of the futurist. Futures projects are political and ethical positions that lead to action.

A project of the future is something that 'can happen' and to accomplish which we take action; it is what Bertrand de Jouvenel and other French futurists have called 'prospective'—the link between science and action. It is to this end that we must rethink futures studies itself: futures studies must become acting for the future.

Futures Thinking as Learning

I wish to stress here that if futures thinking is to direct itself to visioning—searching for and listening to the seed of change—and if it is to build as well, to implement and to act, then it must also be recognised as a learning process. Futurists must learn and help others to learn in future terms. This means learning not only with the aim of preserving the past or perpetuating the present, but learning to anticipate and build the future. It means learning to live with the future—not in the sense of adjusting to it, but in the sense of steering it in the direction we choose. Furthermore, ethics demand that the direction we choose be towards what we consider the best, not merely what will benefit a particular individual or group. This implies the conscious exercise of will and an act of responsibility.

Such a future is directed by what I call 'principles of order as dynamic principles in the flow of life'. As I have explained in another article,[15] these principles in some way govern relations between the microcosm (person) and the macrocosm (the world). We could call them values that undergo mediation from the ontological level. Absolute values exist, in fact, at the ontological level and govern the world; they are outside time but are mediated to the existential level of man in an ongoing process of internalisation. These values at a given historical and spatial moment, and in a continuous and dynamic process, are the dynamic principles of order.

The continuous and dynamic process of learning to understand is the internalisation of values in an existential way. Values undergo constant change in relation to time and space and as such are internalised. It is at this point that learning is important and crucial to futures thinking, and it is this kind of learning that will give human beings the ability to live in a rapidly changing world that seems to overwhelm their capacities to survive and develop mentally and spiritually. We seem to be constantly pursuing technological changes, political changes and economic changes, yet our internal psychological and spiritual structures do not seem able to accommodate such changes. By learning along the lines indicated, internalising values in an existential way, we can learn to survive and also to develop, because in this way we follow principles of order that are outside of time and govern the world but are dynamic in time and space and are made realisable by the human will. In this way, the position typical of our times of fragmentation between the observer and the object, which leads to non-involvement, would be overcome by the mediation of internalised values.

Erich Jantsch[16] cites Abraham Maslow's conclusion that man, by gradual awareness through the well-known hierarchy of needs, attains the capacity for self-actualisation and recognises absolute values as principles of order in the flow of life. According to Jantsch, Maslow's self-actualisation may be understood as the moment in which the identification of subject with object gives us the capacity to recognise the 'order in the flow': the moment of internalised values. Maslow assumes that the role of creativity (primary creativity) enters fully.[17] I, for my part, assert that the role of intuition and spiritual understanding goes far beyond the understanding through reason on which the development of Western civilisation has mainly relied over the last two centuries.

Learning for future-thinking means developing these non-rational capacities, and hence:

1. Learning should be based on awareness of the interrelation between psychological and social character. That is to say, awareness of the interaction between individual temperament and the social character. Each individual has a unique temperament that derives from his or her genetic components, and this develops in dynamics with the social character of the family, the group, the country, etc.
2. Learning should be based on respect for one's own culture and at the same time awareness of other cultures. In a pluralistic world, learning must take account of one's own values, behaviours and lifestyles, and those of others.
3. Learning should be geared to identity; that is, to being oneself, as an individual and as a member of society: to be both an individual continually

changing, and at the same time a dynamic part of society, including a world society. Identity is the capacity to live continuously in a dynamic relation to oneself, to others and to the environment without being completely absorbed by them.

4. Learning must be geared to the ultimate questions of existence: death, love, tragedy, hope, loyalty, power, the meaning of life and the place of the transcendent in human existence.[18] Learning should not avoid raising such questions and should actively encourage the production of tentative answers to them. The aim is not to prove the correctness of any certain set of answers but rather to develop the capability of *perceiving* different answers to such questions; for example, how the conceptions of death, life and love differ in different cultures. Learning understood in these terms is basic to futures thinking and to future studies.

It is through a rethinking of futures thinking and futures studies, within a pluralistic and ethical matrix, that we can ensure that learning overcomes the fracture between the subject and the object. Futurists must think in terms of developing visions, of attaining the capability of searching for and listening to the seeds of change in the process of history, and of building projects for the future through actions based on clearly articulated values, while acknowledging the legitimacy of other perceptions. Futures studies can and must change in these directions so as to become a means for helping human beings better equip themselves to live in a changing world and to steer change to achieve their common benefit.

Notes

1. John McHale, *The Future of the Future*, Braziller, 1969.
2. See also E. Masini, 'The Growth of Futures Research in Global Futures', in Jib Fowles (ed.), *The Handbook of Futures Research*, Westport, 1978.
3. Michel Godet, *The Crisis in Forecasting and the Emergence of the 'Prospective' Approach*, Pergamon, 1979.
4. Fred Polak, *Prognostics*, Elsevier, 1971.
5. A. McDonald, 'Energy in a Finite World' (Executive Summary of the report of the Energy Systems Programme of IIASA, directed by W. Hafele), IIASA, 1981.
6. By 'social system' I here mean 'an interrelated set of social processes in which there is a sufficient feedback to warrant the assumption of some degree of self-maintenance'. See Percy S. Coen, *Modern Social Theory*, Heinemann Educational, 1968.
7. Eric Fromm, *The Sane Society*, Routledge & Kegan Paul, 1956.
8. See E. Masini, 'Children and Development', paper prepared for UNESCO meeting in Tiradentes, Brazil, 1979; and 'Research Project on Education and Youth Unemployment in Italy', prepared for the Ministry of Education, Rome, 1981.
9. Hazel Henderson, *Alternative Futures: The End of Economics*, 1978.
10. Elise Boulding, *Women in the 21st Century*, John Wiley, 1972.

11. See also E. Masini, 'The Role of the Futurist and His Social Responsibility', in *Poland 2000*, Proceedings of the meeting on 'Future Research, Planning, and Decision Making', Jablonna, Poland, 1977.
12. Yehezkel Dror, 'Future Studies, Quo Vadis?', in *Human Futures*, IPC, 1974.
13. M. Botez, M. Celac and P. Dimitriu, *Global Modelling: A Critical Approach*, ICMFDS, 1975.
14. Bertrand de Jouvenel, 'A World to Futurists', in *Human Futures*, IPC, 1974.
15. E. Masini, 'The Quest of Absolute Values: A Reply to Erich Jantsch', *Futures*, October 1976.
16. Erich Jantsch, *Design for Evolution*, Braziller, 1973; and 'The Quest for Absolute Values', *Futures*, December 1975.
17. Abraham Maslow, *The Farther Reaches of Human Nature*, Viking, 1971.
18. Soedjatmoko, 'Cultural Identity of Third World Countries, and the Impact of Modern Communication', in *Communication and Cultural Identity in an Interdependent World*, Proceedings of the World Futures Studies Federation's Sixth World Conference, Egypt, 1978.

4
Reorienting Futures Studies

SOHAIL INAYATULLAH

In the *Future Eaters*, Tim Flannery argues that human beings eat the future.[1] Through our interaction with each other and the environment we reduce the options, the alternative futures for others. This is the central argument of recent critiques of the future as well: the future has been colonised, made into a commodity by corporations, into an official long-range plan by state bureaucrats and domesticated into trivial technological forecasts by popular Western futurists.[2]

In this temporal cannibalism, the main losers are the futures and the presents and pasts of the non-West in all its plurality. Reduced by European political thought to mere nation-states instead of grand historical civilisations, blinded by a history of servitude and obedience instead of creativity and innovation, and epistemologically impoverished believing that it will—hoping to—become a copy of the West, it is not surprising that the non-West sees its future through the eyes and macrohistories of the West. As with all copies poor versions result, and even as technology blurs the distinction between the original and the copy, copying still does not unleash internal creative responses to global challenges, it merely evokes tired mimicry. Whether it is grand history or grand futures, the non-West defines itself through the eyes of others.[3] It is thus not surprising that heroic attempts to consider alternative futures that challenge Western hegemony are seen as threatening to West and non-West. And it is thus not surprising that when the future is imported into the non-West it is done so in its most tame form.

Collusion

This process of importing the future involves a collusion with local actors and agencies. This collusion continues the colonisation of the future. Associations from Pakistan to Malaysia continue to ape Western ideas of the future. Instead of basing the future on internal categories from the civilisational history of Asia, the future is constructed within the modernist framework of strategic and international studies. For example, in Pakistan, borrowing from the dominant discourse strategic studies (with the only real actors being nation-states and their bureaucratic functionaries), the future has become a game of predicting who will win out—will it be China or India, Pakistan or India, Israel or the Arab world?

What results is merely the addition of a temporal dimension to strategic studies. Accompanying this are scenarios focused on Pakistan's alternative economic futures, with economy defined by neo-classical economics. Scenarios focused on the informal women's economy or on bartering between South Asian nations are rarely explored, and when they are, not taken seriously, since they are utopian, that is, not strategic. Thus, although futures studies in the Pakistani case has grown from the non-governmental sector, the notions that inform it are entirely those that fit perfectly into Pakistani governmental space—certainly not the 'game of dissenting visions' that Ashis Nandy seeks to create.[4]

The Western notion of sovereign space is also internalised and used as a framework to envision the future. What results are national studies focused on a particular year such as 2000 or 2020. Malaysia, in particular, has managed to transform the future from an open-ended vision to a slogan for economic development. The future has become a development plan, with the goal that of entry into industrial society. However, as with all proclamations of an official future, there is always dissent. A recent song in Malaysia includes a refrain which goes like this: 'We cannot see 2020, the haze is bad, so is our currency.'[5]

But while humour makes for more interesting futures, a hegemony of knowledge remains wherein the future must be forecast by a national presidential commission, as Ashis Nandy writes, leased out to the appropriate Ministry.[6] The future thus remains technocratic, statecratic and linear. Alternative renderings of time, such as future generations time or those based on cyclical time, are not explored. They are not explored because these alternatives are not conceptually available, since the future is imported as a foreign category. When indigenous forms of time are pursued, they are pursued within the parameters of the parapsychological—as fortune-telling, as astrology. While in itself this is harmless—that is, in the long run these are alternative epistemic systems—they transfer power from the individual to the fortune-teller, creating a class of expertise that disempowers the subject.

The future is thus governmentalised or made fantastic, divined. In both cases, the future as a site of imagination, as creative action, as an attempt to ask what within historical constraints are the conditions for cultural and economic transformation, is lost. The cost of collusion is the colonised self, with the only appropriate future that can get approval derived from Western models of knowledge. This collusion negates the diverse history of the non-West and creates fetish futures based on an idealised version of West (forgetting the West's own history, its technological ascent and its brutal colonisation of others).

Of course there are numerous efforts to break out of these patterns. Clement Chang of the Tamkung University, for example, after twenty years of importing Western notions of the future, has embarked on developing a nativistic futures studies for Taiwan, using but not being used by the external constructions of the future.[7] Tae-Chang Kim has attempted to explore the implications of the paradigmatic concept of *han* (as the paradox of beauty and resentment) for the future of Korea.[8] Puanani and Poka Burgess investigate Hawaiian concepts of time and metaphors of the self in the context of island sovereignty and local community development, creating among other innovations a tourism that involves a sharing of Hawaiian values and also a sharing by the tourist of his or history.

Eliminating the Other

While these efforts to break out of conventional parameters of the future are to be applauded, in general, the future as futures studies, irrespective of civilisation, has itself recently become an exclusive way of thinking and acting. Being a futurist becomes a sign of in/out group, leading some to sign letters and emails as 'futuristically yours'. Futures studies becomes a sort of a cult where the number of those converted to futures studies throughout the world is in itself seen as a sign that things are getting better, as an indicator of a global consciousness change.

This occurs in the West and the non-West. Indeed, in one workshop conducted for an OECD country, I found that members had no agreement on a vision of the future yet felt they were cohesive enough to remain as a group. The process of thinking about the future was enough. The content of the future and the epistemic frameworks that contextualise the many meanings of the future were not important; it was enough to be forward oriented.

The types of critiques of power that derive from this process view of futures studies are those that merely argue that humans are not future oriented enough. While this is quite true—that is, decisions are based on immediate partisan or clan reasons—what is not discussed is why being future oriented is necessarily a good thing. For example, one can articulate numerous reasons why thinking about the future is essential: (1) decisions have long-term consequences; (2) future alternatives imply present choices; (3) forward thinking is preferable to crisis management; (4) further transformations are certain to occur; (5) the future is something we should be concerned with since it has been taken away from us; and (6) unless we create the future it will be created by others.

While at one level these are quite reasonable organising principles futurists are committed to, they are also the platform for the Serbian Socialist Party, which was instrumental in the recent ethnic cleansing in the former Yugoslavia. Nazi leaders would also find these issues unproblematic. Indeed, Wendell Bell argues that the origins of the futures field are partly with the 'social engineering in the early days of communist Russia, fascist Italy and Nazi Germany',[9] of attempts to create ahistorical societies. Certainly thinking about the future is not a sufficient criterion for a good society or effective leadership. Missing are content statements focused on deep democracy (economic democracy and authentic multiculturalism), on ensuring that the Other is included instead of marginalised or as in the Yugoslav case (or that of India *vis-à-vis* Kashmir or Pakistan *vis-à-vis* former East Pakistan) eliminated, and on recognising that there are some deep patterns in history that cannot be eliminated. Perfection, however defined, is not socially possible, cannot be engineered, since the process of creating a perfect society involves the removal of all those who are not quite perfect.

Thus, instead of creating or encouraging different ways of knowing, of deep dissent and deep democracy—of debate as to what the future could and ought to be—what results is futures studies as fetishism. Just as futures studies in the West continues to find ways to justify transnational capital expansion (reducing the risk associated with uncertain futures) and remains tied to Orientalism (fetishising Islamic, Indic and Sinic spirituality), it is not surprising that we find futures studies in nations such as Pakistan or Malaysia continuing Asian hierarchical social structures and the Asian love of conspiracy, that is, the strategic discourse (a typical response of the structurally weak).[10]

The future is believed to be essentially good in itself, forgetting that the future is not empty but full of politics and history, fear and loathing. Merely creating a new category of thinkers called futurists will not in itself change social conditions. Certainly foresight is important but foresight outside of cultural and historical context is meaningless.

In creating a different type of futures studies or approach to thinking about the future much needs to be done, much needs to be deconstructed before a critical traditionalism, a postmodernity not committed to empty ethics or an alternative modernity not based on the particular trajectory of the West, can emerge. What is most important is to create a dialogue of civilisational futures—an authentic conversation of shared ethics—wherein futures emerge that do not romanticise the past (as many culturally driven futures do) or promise a total break with the present (as technological futures do).

Deconstructing Time

To undo the current Western strategic and technocratic hegemony of the future, besides taking a critical approach to what in fact the future is and can be,[11] time needs to be made problematic. For example, instead of remaining only in a linear theory of history, where modernity is the natural goal and others are backward, Aborigines, among others, offer a spiral vision of time. In their view, the present is the dreaming of the ancestors. When we dream—superconscious shamanistic dreaming—we enter the waking state of our ancestors. The future and the past 'snake' back into each other, blurring past, present and future and paradoxically keeping all three alive in the present.[12]

Time must thus be seen not as an independent neutral and ahistorical variable but as a variable that can only be understood in the context of space and observer. The 21st century as the great signature of the future becomes far less imposing when seen from a different view. This is not to say there is no danger in this perspective, after all Pol Pot argued that with the victory of the Khmer Rouge, Time was now beginning—it was Year 0.

However, articulating temporal dimensions that are more focused on cyclical dimensions is much more promising. Even more so would be a macrotheory of change that included linear (progress), cyclical (the rise and fall of civilisations or other units of analysis) and transcendental (that is timeless time, time that is outside our immediate categories, whether achieved through silence in nature, prayer or meditation on the *atman*) approaches.[13] What is of importance here is that shifting to deep historical structures gives a certain weight to the future. The future ceases to be just fantasy; it emerges from a solid understanding of the patterns of change. Macrohistory coupled with the pull of future allows for a more complete view of the future. The future thus constrained avoids technofantasy. However, with the pull of the image of the future, history ceases to be the only variable creating the future—human agency is also active in all its individuals and collective aspirations.

As important as a focus on structure and agency is the location of both within civilisational perspectives of self, time, nature and other. There would certainly be more sanity if a pluralistic, civilisational approach to time was taken. Cults would have a harder time claiming the world was ending since there is always some culture to show that our temporal demarcations are civilisationally idiosyncratic.

While temporal disruptions begin to undo the hegemony of the West, they are certainly not enough; more than contesting the 21st century, and all that it means, is needed.

Outside Hegemony

The West, as has been well argued, has maintained its hegemony by securing: (1) definitional power, defining what is important, what is truth, good and beauty; (2) temporal power, naturalising measurable time; (3) spatial power, transforming sacred spaces into secular spaces, centralising power and exchange in megacities; and (4) creating a centre–periphery-based world economy where wealth trickles upward from the poor to the rich and social control downward from the powerful to those aspiring for power. In contrast to the Aboriginal temporal critique, East Asian nations have been able to challenge the economic power of the West (and to centralise themselves, only keeping the spiritual in ceremonial forms) but have made few inroads into temporal and definitional or epistemic power.

What is needed is a new ordering of knowledge that disturbs, perhaps even makes strange current worldviews, nations, leaders and conventional understandings. Real futures are those that cause cognitive dissonance, that do not make sense to the immediate—not because they are nonsensical but because we do not have the epistemological frames, the language, to comprehend them. But this reordering, these real futures, must go beyond postmodern chic. At heart is not metaphorical playfulness but issues of civilisational life and death, of trauma and transcendence. Café Latte postmodernism, while enthralling, manages to avoid the reality of deep human and natural suffering.

But where can this reordering come from? Using ibn Khaldun's theory of the rise and fall of civilisations, we argue that new models of rationality, temporality and economy will come from those who are outside current systems—West and non-West. In Khaldun's time these were the Bedouins, who gained *asabiya*—the sinews that bind—through struggle with the environment and with the allure of civilisation. Where then are the pockets of dissent, not merely dissent as packaged by current views of polity and economy, but a deep dissent that calls into question the modern episteme?

Alternative futures might thus emerge not from the victor's modernity but from those who accept it as part of the many. One cannot be multicultural if one is in a dominant situation, if one has levels of power over others. It is he or she who exists on the margins, who must learn different ways of knowing to survive, who can know differently (through trauma and transcendence) and can thus offer alternatives beyond postmodernity and the information highway.

Alternative futures in general, then, are more likely to come from the non-West or beyond-West, from the indigenous traditions: Tantra, the Aboriginal, the Hawaiian, the African and the Muslim (not the Muslims who live in an oppositional relationship to the West, but Muslims who dissent from mullahist Islam and secular nation-statism). Others marginalised by the victory of world capitalism, women, children, the disabled and of course men, individually critical of the system they structurally perpetuate, too can offer alternatives.[14] For Riane Eisler, for example, it is women—in league with progressive men—who will create alternative futures, ending 5,000 years of patriarchy. They, she argues, have the potential to create more partnership-oriented futures committed to greener values.[15]

The future thus can come from the non-West, for even if it has partly entered scientific and business time, it has not lost its historical ecological diversity of time and culture. For example, even under the veneer of accepting Western notions of death—that death must be battled through technology—the non-West retains varied cultures of death. It sees death as the meaning of life, to be mystically transcended (the *Upanishads* for example), as resignation (karma, Allah's will) and as joining one's ancestors. But for the West fear of death is the defining impulse. 21C[16] perhaps is a civilisational boast, a call out to the universe: we have made it! 2,000 years and still going strong—Nazism defeated, fascism defeated, communism defeated, and now only Islam, uppity East Asians and nature left (Africa will take care of itself and South America has already joined). Still, there are 1,000 years for that project. Most likely we will only need 100; after all, all roads lead to the melting pot.

However, while the marginal can offer local responses to globally shaped futures, since they are focused on defending their cultural territory, they are often less able to offer responses that are both global and local. They present cyclical responses but not evolutionary spiral models of the future wherein the past is not essentialised and capital/technology is not seen as deterministic. The past in these local responses, as with Gandhi, is seen as the future, instead of the evolutionary link to an alternative future that includes yet transcends the past.

Sarkar as Epistemic Transformer

One example of an evolutionary response comes from Indian activist and philosopher P. R. Sarkar. Sarkar's work is important to us for many reasons. Not only did he redefine rationality, seeing it in spiritual and social justice terms, but he placed the subtlety of inner love at the centre of his cosmology. But while love was the base, he did not neglect the harsh realities of the world-system. While certainly his work can be seen as part of the larger global project of creating a strong civil society to counter the waves of corporatist globalism, his movements are unique in that while most social movements are Western, highly participatory, goal oriented, short term, and single-issue based, Sarkar's offer a genuine non-Western future.[17] His movements are:

1. Third World oriented, hoping to be the carriers of the oppressed yet also seeing the oppressors in humanist terms;
2. Tantric, focused on reinvigorating mystical culture and not necessarily on immediate efficiency;
3. comprehensive, working on many issues (and not just on the issue of the day) from women's rights and workers' rights to the prevention of cruelty to animals and plants;
4. oriented to the very long term, hundreds of years, that is, structures and processes that cannot fulfil their goals for generations ahead;
5. committed to leadership-creation and not just organisational development, thus avoiding the bureaucratic tendency; and
6. trans-state oriented, not solely concerned with nation-states and ego-power but acknowledging that there are four conventional types of power—worker, warrior, intellectual, economic—and that the challenge is to develop processes that create a fifth that can balance these forces.

Underneath these projects has been Sarkar's effort at creating and using a new language (*samaj*, *prama*, *microvita*, *samadhi*, *sadhana*) and new metaphors (Shiva dancing between life and death) to help as the vehicles of the good society he envisioned.[18] But it was not perfection Sarkar was after. Influenced by Indian thought, he understood that there are deep evolutionary structures that cannot be changed, but that certainly the periods of exploitation can be minimised. Perfection for Sarkar was only possible for individuals in the spiritual inner sphere, timeless time.

Sarkar's perspective, as well as the many other non-Western perspectives, hopes to create not just a global civil society, as with normal Western social movements, but what Ashis Nandy has called a gaia of cultures.[19] Civil society is a response to particular dynamics of European history with its division between religion and Church, state and society. For Muslims the dream is a global *ummah* (a world community or *ohana* or family). For the Maori, writes activist Ramana Williams, the appropriate term is the creation of a *whanau*. 'It means a vast universal family that connects the stars and the moon, the earth, and the sky and all life forms that reside therein, the world of animation and inanimation, the worlds of the living and the dead. It is not *universum nullium* but knowing and intelligent, and linked by genealogy.'[20]

While Sarkar is my preferred among the most important of non-Western visionary thinkers and activists (since his work while ideational is also firmly rooted in the practice of economic and social movements), he is certainly not alone in his articulation of a different universal family. Muslim thinkers have attempted to create a new discourse in which to Islamicise science, knowledge and the future, with the hope that Islam can be rescued from both mullahs and secularists.[21]

In my work in this area, I have deconstructed traditional models of forecasting —for example, simulation models from the Club of Rome and others—to show how their authors define variables in their model from their own ideological positions. Even as they claimed neutrality, their forecasting models reinscribed their particular managerial scientific view of economy and society as well as their cultural assumptions towards population and wealth growth.

Ashis Nandy has reminded us that the great example of a successful futures studies, that of John Kennedy promising that a man would land on the moon in *x* years, should be seen not as a victory for visioning and idealism but as a victory for scientific economism, for it was this social organisation that was buttressed.[22] The failure of economic development as a vision gives testament to this; that is, when culture or history is involved it is far more difficult for managerial rationality to be hegemonic.

These interventions intend to create a different type of futures studies; they disrupt conventional views of self and future—showing them to be ideological—and thus create the space for other, alternative futures. Among the spaces created is what Manas Ray following Partha Chatterjee calls 'alternative modernities'. This future does not require a return to a traditional religious ethos; rather, modernity—meaning the challenge to feudalism and religious dogma—is embraced, but embraced in historically unique ways, and not as an imitative response to the West.

Future-Generations Thinking

Much of the critique of futures studies has recently come through future-generations thinking. It is focused on family, intergenerational links, Confucian and Buddhist thought, the repeatability of history, a spiritual and collective approach to reality, and decision-making and pedagogy that focus on enhancing wisdom.[23] This is a dramatic move from Western futures studies. For example, a spiritual and collective view of individual choice and rationality is that choice is contoured by both the *aina* (land, as in the Hawaiian tradition) and the heavens.

Rationality is not individual nor instrumentally based but collectively linked to *samaj* (Sarkar's Tantric Indian idea of a society/family moving together towards an integrated and balanced society) and it is given by God. Rationality is not merely logic but inclusive of other ways of knowing such as intuition, the voices of the spirits/ancestors and the altered fields of awareness generated by interaction with the wildness of nature. Rationality is thus tied metaphorically speaking to the heart, to others and to nature—ultimately rationality is about understanding that that is eternal. Finally, future-generations thinking is global; it is a view that while future-generations thinking is civilisationally based, its message is universal, searching for similarities among the many differences between peoples, creating a family of civilisations.

However, the critical and civilisational perspectives outlined above are obviously not a dominant perspective in futures studies. Western futures studies dominate. Two important reference sources show evidence of this. The recently released *Encyclopedia of the Future* specifically goes out of its way to exclude non-Western references since these books will be difficult for Americans to find.[24] Referencing, however, is not merely a sign of a bookstore but symbolic of what is important. The knowledge categories of the classification scheme in addition only make sense to the Western worldview—that is to say, the choices of what is covered: Italy and India but not Pakistan or Malaysia; cross-impact analysis and Delphi but not the I-Ching; Hinduism but not Tantra. Of the classical thinkers included there is no mention of ibn Khaldun, al-Baruni, Ssu-Ma Chien or Confucius. Crucial terms such as karma, *prama* (dynamic balance, the central principle to a society that is in equipoise between the spiritual and the material, the inner and the outer, struggle and acceptance), *mana* (the life force), *aina* (land but not real estate), and *ilm* (knowledge), which are central to understanding the epistemologies and methods of other cultures, are missing. While the *Encyclopedia* is a good initial effort—and its editors should get some credit—its definitional power is disastrous for those wishing for a future different from the present.

Of course, one cannot cover everything; but the omission is predictable, since the non-West is there to be futurised not to provide authentic alternatives to the global problematique. This is largely because futures studies is spatially only committed to forward time; missing are layers of hierarchy, methods that analyse the layers of causation,[25] what Richard Slaughter has called paradigmatic methods.[26] For example, most futures studies stays focused on the litany level of problems. These problems, such as the environment, population, crime, terrorism and protectionism remain unlinked and the historical reasons behind them unquestioned. Moreover, the worldviews underneath these problems are not unpacked. For example, overpopulation is seen as problem because it leads to more poverty. What is forgotten is that overpopulation is a symptom of an insecure future, and most importantly what is forgotten is women's power and that some cultures, in fact, love children. The deeper definitions behind issues are also not explored; thus, unemployment is a problem within the current social organisation of defining distribution and wealth through jobs. Employment, like the category of population, is recent and will most likely disappear if we enter post-capitalist societies.

The recent contribution by Edward Tenner to the *Britannica 1998 Yearbook of Science and the Future*[27]—while a solid review of futures studies—makes the same mistake of equating thinking of the future with seminal works from the European Enlightenment. This linear history of the future again limits the richness of the future. It assumes the future is merely about what technologies will be present and the ideas of progress associated with them. This is certainly one layer of the thinking about the future, but as important are the social and political factors and the paradigms and worldviews that provide the base for the technological. Nor does Tenner present examples of utopian thinking in other cultures. However, Tenner does rightly understand that futures studies will rise and fall not because of its predictions but because it offers a richer debate on questions concerning the good society and self. He writes, 'At its best it [futures studies] offers not a crystal ball, but a kaleidoscope.'[28]

This, of course, is not a critique of the authors but a pointing-out that these images are unavailable to them: fish certainly did not 'discover' water and the West will not find the future even as it predicts it. For example, the one scenario that American Second World War planners did not plan for was attacks by Japanese kamikaze pilots.[29] However, anyone remotely sensitive to East Asian culture would understand the different meanings given to the giving of life for a higher cause—self-immolation and other activities occupy less controversial space in large parts of the non-West.

Needed, then, are multicultural futures and metaphors that can embody these differences as well as critical methods that clearly show them. Equally important is a macrohistorical base to the pull of the image of the future as well as a civilisational dialogue on desirable and undesirable futures.

What is fortunate about futures studies is that these alternative methods, processes and visions are there and are flourishing. The lack of a hegemonic organising variable within the field allows for these differences. While many are merely fetish, aping other disciplines or the West, there are seeds of transformation throughout the planet that give hope for a different future, for different futures. Let us hope these seeds can sprout and grow before the future-eaters find them.

Notes

1. Tim Flannery, *The Future Eaters*, Kew, Victoria: Reed Books, 1994.
2. See Ziauddin Sardar, 'Colonising the Future: The "Other" Dimension in Futures Studies', *Futures*, 25(2): 179–87, 1993. Also see Jim Dator, 'De-Colonizing the Future', Ontario Educational Communications Authority, April 1975; Sohail Inayatullah, 'Colluding and Colliding with the Orientalists', *Futures*, 25(2): 190–5, 1993; Ziauddin Sardar, 'The Problem', *Seminar*, 460: 12–19, December 1997.

3. Zia Sardar, Ashis Nandy and Merryl Davies, *Barbaric Others: A Manifesto on Western Racism*, London: Pluto Press, 1993.
4. Ashis Nandy, 'Bearing Witness to the Future', *Futures*, 28(6–7): 636–9, 1996.
5. Virginia Moncrief, 'Mahathir, Wounded Tiger', ABC Television Documentary, 1998.
6. Nandy, 'Bearing Witness to the Future', p. 639.
7. Clement Chang, 'Three Decades of Futures Studies at Tamkang University', in Sohail Inayatullah (ed.), *The Knowledge Base of Futures Studies*, vol. 4, Melbourne: Futures Study Centre, 1998. The first three volumes were edited by Richard Slaughter and published by DDM, 1996.
8. Tae-Chang Kim, 'Coherence and Chaos in Our Uncommon Futures—A Han Philosophical Perspective', in Mika Mannermaa, Sohail Inayatullah and Richard Slaughter (eds.), *Coherence and Chaos*, pp. 41–8, Turku: Finland Futures Research Centre, 1993.
9. Wendell Bell, 'The Sociology of the Future and the Future of Sociology', *Sociological Perspective*, 39(1): 40, 1996.
10. Sohail Inayatullah, 'Imagining an Alternative Politics of Knowledge: Subverting the Hegemony of International Relations Theory in Pakistan', *Contemporary South Asia*, 7(1): 27–42, 1998.
11. Sohail Inayatullah, 'Deconstructing and Reconstructing the Future', *Futures*, 22(2): 115–41, 1992. Also see Richard Slaughter, *Futures: Concepts and Powerful Ideas*, Melbourne: Futures Study Centre, 1996.
12. Paul Wildman, 'Dreamtime Myth: History as Future', *New Renaissance*, 7(1): 16–20, 1997.
13. Johan Galtung and Sohail Inayatullah (eds.), *Macrohistory and Macrohistorians*, Westport, CT and London: Praeger, 1997.
14. At the same time there is no guarantee of this. Once those in the periphery enter the gates of the centre they have a habit of becoming worse than the original exploiters.
15. Riane Eisler, *Sacred Pleasure*, San Francisco, CA: HarperCollins, 1996.
16. Peter Lorie, *The Millennium Planner*, London: Boxtree, 1995.
17. P. R. Sarkar, *PROUT in a Nutshell*, vols. 1–25. Calcutta: AM Publications, 1988. Also see Sohail Inayatullah, 'Sarkar's Spiritual Dialectics', *Futures*, 20(1): 54–65, 1988.
18. Sohail Inayatullah and Jenny Fitzgerald (eds.), *Transcending Boundaries: P. R. Sarkar's Theories of Individual and Social Transformation*, Brisbane: Gurkul Press, 1998.
19. Ashis Nandy and Giri Deshingkar, 'The Futures of Cultures: An Asian Perspective', in Eleonora Masini and Yogesh Atal (eds.), *The Futures of Asian Cultures*, Bangkok: Unesco, 1993.
20. Ramana (Ross) Williams, 'Communication Paradigms and the Challenge and the Other: A Trialogue between Western, Tantric and Indigenous Traditions', dissertation honours, the Communication Centre, Queensland University of Technology, 1996, p. 42.
21. Zia Sardar, *Islamic Futures: The Shape of Ideas To Come*, London: Mansell, 1995.
22. Ashis Nandy, *Science, Hegemony and Violence*, Delhi: Oxford University Press, 1993.
23. Richard Slaughter and Allen Tough (eds.), 'Learning and Teaching about Future Generations', special issue of *Futures*, 29(8), October 1997).
24. George T. Kurian and Graham T. T. Molitor (eds.), *Encyclopedia of the Future*, 2 vols., New York: Macmillan Library Reference, 1996.
25. Sohail Inayatullah, 'Causal Layered Analysis: Poststructuralism as Method', *Futures*, forthcoming, 1998.

26. Richard Slaughter, 'Developing and Applying Strategic Foresight', *The ABN Report*, 5(10): 7–15, December 1997.
27. Edward Tenner, 'To Glimpse Tomorrow', *Encyclopedia Britannica Yearbook of Science and the Future*, 1998: 56–71.
28. Ibid. 71.
29. Ibid. 69.

5
Feminising Futures Studies

IVANA MILOJEVIC

The majority of the liberal, or 'progressive' futurists today acknowledge the fact that futures studies—a not yet recognised field of enquiry within traditional disciplinary scientific divisions—has been dominated by one civilisational view of time, reality and space. The futures of non-Western people and countries have been colonised in a similar way to their presents and pasts.[1] But even among the most progressive futurists there is a very strong underlying belief that futures studies is at least gender-free. These futurists believe that futures studies is the field in which personal values and attributes transcend polarised gender divisions. Some of them would rather belong to a 'people's movement' than to one that is part of and belongs to a particular gender group; or they describe the future like a toilet with separate entries but with the inside the same for everyone.

This reminds me of the debates and realities in my own country, and the efforts to transcend particular national identities while creating a new one, Yugoslav. Not surprisingly, it was always easier for the largest national group within the former Yugoslavia, the Serbs, to have their identity changed, as they did not feel that this new identity would deny their previous one. On the other hand, marginal national groups, not just in Yugoslavia, often see the overlapping globalising identities as a threat to their own, as they realise they will always be outnumbered. The reason why I, and some other women futurists, believe we should still occasionally work within 'women's groups' is because within futures studies—especially where money and status are involved—women are outrageously outnumbered. The umbrella of futures studies should be big enough to cover everyone's issues and concerns, but in reality, the famous futures fork is always leaning towards the male side and masculinity.

What is even more disturbing is the fact that most women futurists within 'people's movements' work within accepted styles, on problems and issues as defined by masculinist concerns. This is, again, not surprising. Past and even present events teach us that if women 'come out' as feminist, or try to discuss women's own views on the future, they usually come under vicious attack. One example is a special report in *The Futurist* on 'Women's Preferred Futures'.[2] This report was initially included in the journal as a result of women futurists' complaints that an article in the journal on 'Women of the Future: Alternative Scenarios' had been written by a man.[3] Women futurists who sent the letter, Hazel Henderson, Eleonora

Masini and Riane Eisler, did not want to 'condemn' the article itself, believing it was 'well meaning', but felt that women futurists should have been allowed 'to speak for themselves'.[4] This feeling was intensified partly because of one illustration on the same page representing a chained woman.

Behind all the immediate and transparent reasons, the reaction was probably also partly intensified as a result of long-term frustration with male domination in the field. Not only are men the greatest experts when it comes to the future in general, or when it comes to every particular aspect of it, but their views and opinions are also consulted when it comes to women's futures, issues and concerns. In response to critiques of the representation of women, the World Future Society, which publishes *The Futurist*, decided to 'put up' with women's issues, and invited women futurists to 'tell their vision of a preferred future'.[5] The section was 'written, edited, typeset, designed, and illustrated solely by women'.[6] Not long after, this special report came under attack in the letters section. Even though this report asked women futurists what would be their *preferred* vision for the future, women who contributed were labelled as an 'unrealistic bunch'.[7]

The other critique, also by a man, is a paradigmatic critique that has followed feminism from its early days: this bunch could not claim to represent the 'majority of women' and instead the average woman should have been asked to 'speak for herself'.[8] While it is, of course, perfectly acceptable that Western male futurists can make any generalisation or universalistic statements about 'the future', when it comes to women futurists' visions, 'their opinions and prophecies' are labelled as 'self-serving of their own emotional and financial needs'.[9] The writer of the letter suggested that we should instead try to go out and find the average woman, meaning a 'mother, homemaker, wife, school volunteer, factory or office worker'.[10] The only letter sent by a woman, however, labelled one particular aspect of the report as 'enriching', as it gave an alternative to the issue she, in her working life, finds 'distressing'.[11]

For most gender-conscious women futurists it is obvious that there is a big discrepancy in the way most people think about future trends and their alternatives, depending on which gendered interests they represent. Feminine alternatives are usually labelled as poor writing, or naive, or without enough substance, or utopian, while masculinist images, especially techno-maniacal and dystopian, are usually seen as realistic, far-reaching and logical. It is interesting that especially the darkest images of the future get to be chosen as 'realistic'—somehow, people 'take it as axiomatic that fears are realistic and hopes unrealistic'.[12] For feminist futurists it is also obvious that the way to the 'future's toilet' is all high-tech, making-life-easier, on the gentlemen's side, and far too difficult, naturalised with thorns and bushes, on the ladies' side.

The domination of the masculinist images of the future has now reached a new peak. These images are accepted by globalising popular media, by local and global policy-planners and even by many liberal futurists. They all give priority and attach a higher value to grand historical analyses and issues, concentrating especially on discussions of where power is going next. This is where women futurists might rather wish to be on the 'other side', either among 'average women' or among radical feminist separatist groups. Because, unfortunately, power in the 'next millennia' does not seem to be going in the direction of women. Just take the year 2200 as an example: according to Kurian and Molitor it will be an era in which women own up to 20 per cent of the world's property (a dramatic increase from the hardly believable 1 per cent, as it is apparently today).[13] At the same time, world income received by women will increase from the current 10 per cent to 40 per cent, which would represent a significant increase—if it is realised.[14] Kurian and Molitor, however, do not state on which 'facts' they base their forecasts. In fact, there is an ever increasing gap between rich and poor, and women are, unfortunately, still the majority of the world's poor.

Postmodernism and the influence of non-Western feminism have changed the way we write and think about 'women' and destabilised the previous universalistic conception. We now accept that the category 'women' is as diverse and different as the category 'men' or 'people', since there are certain things we, as people, all share. However, there are also certain things we, as women, still have in common. One of those things is that we (women) all lack the most important resources for liberating ourselves and the future from masculinist domination: resources in time and personal energy. Our time and our energy are shattered over the multiplicity of tasks necessary for adjustment and survival within patriarchal societies. Furthermore, together with many other marginal groups we lack the initial resources in wealth, education and knowledge, informal networks and even more importantly the will to engage in the power battle.

I will now further explore women's tradition of thinking about and influencing the future, and contemplate how the future could be liberated or de-masculinised.

Women and the Future

At present, the reality is that women are not in charge of the future. Despite being 'practising' futurists'[15] women do not decide much about the general future, nor are they expected to. But that was not always so. The importance of looking in the past for our efforts in thinking about and creating of the future can be summarised in a famous sentence by Kenneth Boulding: 'If it exists, it is possible.'[16] So, even if present trends do not promise much to girls and women of the future, our own ability also to create the future certainly gives us more hope.

The past

The evidence of women's one-time importance when it comes to understanding and creating the future can be easily found in the realm of old and long memories —those expressed in Slav, Greek, Roman, Nordic, Saxon and Indian mythology. In my own Slav tradition, there are stories of the so-called *sudjenice* (from the Serbo-Croatian word for destiny, *sudbina*), who are represented as three women in charge of deciding everyone's personal destiny. They are also known as *sudjaje*, *rodjenice*, or *rozanice*.[17] They arrive when a child is born and decide every particular aspect of her/his future life. Their will can not be changed, but people can try to please them and in that way increase the chances of a positive outcome.

In the Greek tradition, they are the Fates, or *Moirae* ('cutters-off', 'allotters'), who personify the inescapable destiny of man. Clotho, the spinner, spins the thread at the beginning of one's life; Atropos, the measurer, weaves thread into the fabric of one's actions; and Lachesis, the cutter, snips the thread at the conclusion of one's life.[18] The process is absolutely unalterable, and gods as well as women and men have to submit to it.[19] As goddesses of fate, the *Moirae* 'necessarily knew the future and therefore were regarded as prophetic deities: thus their ministers were all the soothsayers and oracles'.[20] The Roman equivalent were *Fortunae*, or (apparently in the medieval period) three *Parcae* ('those who bring forth the child'): Nona, Decuma and Morta. Most religious traditions call the Fates 'weavers' and the Latin word *destino* means 'that which is woven'.[21] In the Nordic tradition they are called *Norns*. There are also three *Norns*: Urd, representing fate, Verdandi, representing being, and Skuld, representing necessity. The *Norns* could change into swans for ease of travel but they were usually to be found near the roots of the ash tree Yggdrasil. The Yggdrasil tree had huge roots: one stretched to the underground spring of Urd (earth); the second reached to the well of Mimir, the well that was the source of all wisdom; and the third went to Niflheim, the underworld presided over by the goddess Hel.[22] Each one of three Norns knows and is accredited with a particular province: Urd knows the past, Verdandi the present, and Skuld the future. In fact, it seems that the only deity who was especially in charge of the future was not a deity but a deitess, Skuld. According to Barbara Walker, all of Scandinavia and also Scotland was named after Skuld, or 'as Saxons called her, *Skad*'.[23] The Saxon *weird sisters* also represented the past, present and future: become, becoming, and shall be.[24] It seems that *Norns* and their equivalents were based on the great Indo-European Goddess as Creator, Preserver and Destroyer and are in some ways close to the Indian goddess Kali.[25] Kali also symbolises 'eternal time and hence she both gives life and destroys it'.[26] Mother Kali continually ruled the Wheel of Time (*Kalacakra*), where all the life-breath of the world was fixed.[27] In most archaic traditions, 'the deciding of men's fates was a function of the Goddess'.[28] Goddesses were also often creators of the universe: for example, in Sumerian cosmogony the ultimate origin of all things was the primeval sea personified as the goddess Nammu—the goddess who gave birth to the male sky god, An, and the female earth goddess, Ki.[29]

The past and the present

In patriarchal times the Fates became 'witches': Shakespeare's three witches were called weird sisters (adapted from the Saxon tradition).[30] The Christian Church appropriated this ancient belief and transformed the trinity of She-Who-Was,

She-Who-Is and She-Who-Will-Be into its holy trinity of the Father, the Son and the Holy Spirit.[31] As God became male so did time, so did the future. Men decided which parts of our past tradition deserved to be recorded[32] and passed on to future generations; they decided which direction we should choose next. From many secret symbols that celebrated the power of women and female principles, the symbol of Venus (representing love and sexuality) was chosen for women. If we try to deconstruct this symbol we can see that its essence is in the cross below, the cross that, especially if surrounded with the circle, has traditionally been the symbol for the Earth. Men's symbol, the sign of Mars (god of war), has its essence in the arrow: a symbol often viewed as a phallic symbol, as a weapon of war. In the male symbol the arrow points towards the upright direction, which is also, not surprisingly, how we draw trends and movements towards the future. The present understanding of women is in their role as conservers, deeply rooted in the ground, with their essence in the body. Men are the ones who transcend their mind and are in charge of the future, as they are the ones who bring about political changes and preach radically new prophecies.

I said it is not surprising that we draw future trends in the same way we draw the symbol for the god of war as this is exactly the direction we are heading towards. Each year we face more and more people being killed, especially civilians in wars between countries, and in wars on the streets. We are fighting against 'Mother Nature' and against our own, inevitably animal bodies. Our most popular images of the future are the ones of war-games, of the future with ever more powerful weapons and ever more powerful enemies. Conquest in the future is as important as conquest now, and it is both the ultimate conquest of old enemies and the battle for life and death with new ones (aliens, cyborgs, mutants, androgynes). This has resulted in the sad fact that, according to a recent Unesco study, the killer robot played by Arnold Schwarzenegger, or the 'Terminator', 'was the most popular character among the world's children'.[33] The survey, which was billed as the first worldwide study of violence in the media, said 88 per cent of children around the world knew the Terminator, who was 'a global icon' and that more than half the children—raised in environments of violence—wanted to be like him.[34]

The recent present and the future

Such an idiotic obsession with death, killing and self-destruction has had the impact of awakening worshippers of peace, nature and tranquillity. If Riane Eisler is right, time is ripe for yet another shift in the power battle between the female and the male principle. Women, say Aburdene and Naisbitt, have lately evolved 'into a more complex state of wholeness', successfully absorbing positive masculine traits, and will lead the way to the future.[35]

As a part of this process many feminists have tried to revive the Goddess as a symbol of this power shift. The reason behind the Goddess reawakening is empowerment, for as 'long as people visualize God as male, women are diminished and inferior'.[36] But this time it might be much more difficult for the Goddess to express her female principle. For postmodernists, essence as 'women' (or female) and 'men' (or male) does not exist as such any more. In fact there are hardly any criteria left that would suffice to describe two different and opposite genders. Criteria such as appearance can be challenged by transdressers and transvestites. Sexual orientation has always been problematic as a criterion since homosexuality among humans has (probably) always been present. Thanks to modern medical science, the natural characteristics of the sexes can be transformed and changed, women becoming men and vice versa. Woman (or man) as a social category is also problematic since any universalist statement about woman (man) can be questioned from the position of epistemological (and group) minorities and different perspectives. The Reawakened Goddess of the Future will have to work rather in a context of multiple-gender diversities of the future than in the context of the traditional female–male polarity.

But this is not the only challenge the awakened Goddess is facing. She ruled in societies that belonged to a totally different historical context. The renewed symbols of goddesses are also symbols that make much more sense within the context of agricultural societies. The cyclical understanding of time, reclaimed as women's, as opposed to a linear patriarchal one, probably resulted from observations about cyclical changes within nature—observations obviously extremely important for agricultural societies. It is difficult to revive the ancient cults of earth and goddess worship in times when fewer and fewer women live by the dictates of their own natural cycles, when an enormous number of the world's women live in cities, and when reproduction within women's bodies might soon become obsolete—several thousands years of masculinist rites and gods notwithstanding. Donna Haraway senses this change while declaring that she would rather be a cyborg then a goddess.[37]

And our own *Norn* Skuld does not sit under the secret ash tree any more, but in front of the computer, with her sister Urd.[38] While surfing the Net we can visit 'The Sacred Shrine of Skuld-sama', where we are welcomed to an information resource and place of worship dedicated to Skuld, the technologically minded young goddess from 'Aa! Megamisama'. The Skuld of today is '12 Earth years old, 150 cm tall, with brown eyes and black hair, while her vital measurements are se-cr-et! She is a second class goddess with limited license.'[39] Her domain is still the future and her travel medium these days is water. We are also informed that she likes her older sister Belldandy—and '131's Ice Cream'. Besides eating ice cream her favourite activity is to build all sorts of mechanical devices. Her best inventions include Banpei-kun, the anti-Marller defence robot and Skuld's Own 'Debugging Machine, a modified rice cooker that specializes in catching bugs in a manner similar to the Ghostbusters' Ghost Trap. [She is still] a very strong-willed girl [displaying] sometimes fiery temper, [and] is in charge of "debugging" the Yggdrasil mainframe up in the Heavens, as well as the occasional bug that appears on the surface world'.[40] She has her own image, music and sound, literature and movie world library, her own desktop themes (Skuld backdrops, cursors, a game, and more!) and, of course, her own mailing list.

Women as practising futurists

However, it is not only in the distant past or in the emerging future that women thought and think about or tried and try to influence the future. Even during the peak of patriarchy there were some individual women who were trying to change gender relationships. At least, women have always been 'practising futurists'. And they have always been active within grass-root movements. At the same time, though, women did not and do not decide much about the general future. Women's encounter with the future is reserved to better care for future generations and present households. Therefore, women have to know something about the future, but not too much. They should organise local networks to support global political and economical processes, but should not intervene in the essence of the latter. Even old and traditional women's activities directed towards influencing the future (through their role of witches or fates) were primarily local, personal, family- and community-oriented.

The feminist dictum of the personal being political suddenly gave us the legitimation to bring what has always been extremely important to us (personal relationships, family, community) to the societal level. For example, the issue of violence against women is less and less considered a private matter, an event that happens and should remain behind closed doors. Rather, it is seen as a global issue; and actions in the prevention and reduction of violence are therefore being conducted at the world level as well.

The legitimation of 'women's issues' has created the opportunity for many women futurists to write about not only local but also global futures directions. Many are envisioning radically different future societies and suggesting feminist (or women's) alternatives to patriarchy. Their images can easily been labelled as utopian: for example, Boulding's vision of gentle/androgynous society or Eisler's partnership model/gylany. However, the images brought to us by the work of Boulding, Eisler and feminist fiction writers, utopian or feasible, are extremely important for the de-masculinisation of the future, because what we can imagine, we can create.

Elise Boulding, Riane Eisler and feminist utopias

Elise Boulding's image of the 'gentle society' is an image of a society situated within a decentralised (and demilitarised) but still interconnected and interdependent world.[41] The creators of the gentle society will be androgynous human beings (she brings examples from history in the images of Jesus, Buddha and Shiva), people who combine qualities of gentleness and assertiveness in ways that fit neither the typical male nor the typical female roles. The coming of the gentle society will, according to Boulding, happen through three main leverage points: family, early-childhood school setting (nursery school and early elementary school) and community. Boulding believes that both women fiction writers and 'ordinary' women imagine and work in a direction of creating a more localised society, where technology will be used in a sophisticated and careful way to ensure humanised, interactive, nurturing and non-bureaucratic societies. Through women's triple role of breeder–feeder–producer women can bring radically different imaging and are therefore crucial for the creation of a more sustainable and peaceful world.

For Riane Eisler—in our nuclear/electronic/biochemical age—transformation towards a partnership society is absolutely crucial for the survival of our species.[42] Since today, owing to many technological changes, our species possesses technologies as powerful as the processes of nature, if we do not wish to destroy all life on this planet we have to change the dominator (patriarchal) cultural cognitive maps. In gylany (as opposed to androcracy) linking instead of ranking is the primary organisational principle. It lacks institutionalisation and idealisation of violence and stereotypes of masculinity and femininity. More equal partnerships exist between women and men in both the so-called private and public spheres and there is a more generally democratic political and economic structure. Eisler also envisions gylany as society in which stereotypical 'feminine' values can be fully integrated into the operational system of social guidance.

Boulding's and Eisler's images of future societies correspond in many ways to feminist fiction writings. Their writings also correspond to most grass-roots women's activities and to women's involvement within the peace or Green social movements; and they are obviously more similar in spirit to the ancient goddess Skuld than to her recent technological rebirth on the World Wide Web. For Boulding, education is one of the most important social institutions, crucial for our future. Similarly, in most feminist utopias, education and motherhood are not only extremely respected, but are sometimes also the main *raison d'être* of the utopian society in question. There are also some other common themes in feminist utopias: future societies tend to live in 'peace' with nature and have some sort of sustainable growth; they are generally less violent than present ones; families seldom take a nuclear form but are more extended (often including relatives and friends); communal life is highly valued; societies are rarely totalitarian; oppressive and omnipotent governmental and bureaucratic control is usually absent, while imagined societies tend to be either 'anarchical' or communally managed.[43]

On the other hand, the masculinist colonisation of the future brings about images of totalitarian future societies, societies with some sort of feudal social organisation, and where 'progress' is defined in terms of technological developments. Feminist writings about the future might be 'naive' or too utopian, but mainstream images are rather evil and dangerous. Some of the elements within feminist imaging of the future are rather reminiscent of the times when gender relationships were more equal—in past agricultural and matrilocal societies. But even with all the recent technological developments there is nothing in the world (except our patriarchal cultural cognitive maps) to prevent us from giving priority to education and parenting instead of to the corporate and military sectors. We can use new technologies to repair environmental damage rather than to keep on increasing it. We can use them to improve the health and happiness of future generations rather than to steal the future from them. New technologies can also help create a system of direct democracy or connections between world government and local communities. The Net can enable equal access to social groups previously discriminated against because of their disability, gender or race. It can help celebrate, understand and learn about diversity, difference and 'the Other' rather then making our songs unison.

The De-masculinisation of Futures Studies

If futures studies opted to work within 'feminine' guiding principles it would most likely prioritise the futures of education, parenting, community, relationships and health—the real grand issues! The futures studies methods most commonly used would not be forecasting or trend analyses but backcasting as well as visioning workshops with disadvantaged groups, in order to empower them. Futures research would always have gender differences in mind, from deciding which problems are going to be investigated to research design and collection and interpretation of data. Futures research would not only acknowledge the pervasive influence of gender but would also be concerned with its ethical implications.[44]

Sometimes it is quite easy to make the necessary changes. For example, a sentence such as 'a host of new fertility treatments now enable barren women to have a much-wanted child'[45] should read 'a host of new fertility treatments now enable childless couples to have a much-wanted child'. The language in the first sentence is that of patriarchy, where it is always women who are blamed for the lack of children in marriage and where the responsibility for child-bearing and -rearing is solely women's. The second sentence is more in accordance with current knowledge in medicine about the causes and reasons behind infertility—men's inability to father a child being equally the cause of the problem. It is also the language of potentially emerging egalitarian relationships between genders and societies, where parenting and education of children are going to be respected more—both by men and by general society.

The de-masculinisation of the future and futures studies seems very radical and most likely it will be a rather slow and difficult process. But the change needed is no more radical than the change that transformed weird sisters into witches, triple Goddess into Holy Trinity, and Verdandi into Belldandy. The emerging change might be utopian, but it is possible.

Notes

1. Zia Sardar, 'The Problem', *Seminar*, 460: 12–19, December 1997; Sohail Inayatullah, 'Listening to non-Western Perspectives', in David Hicks and Richard Slaughter (eds.), *World Yearbook of Education 1998*, London: Kogan Page, 1998, pp. 55–69.
2. *The Futurist*, 31(3): 27–39, May–June 1997.
3. *The Futurist*, 30(3): 34–8, May–June 1996.
4. *The Futurist*, 30(5): 59, September–October 1996.
5. Ibid.
6. *The Futurist*, 31(3): 27–39, May–June 1997.
7. *The Futurist*, 31(5): 2 September–October 1997.
8. Ibid.
9. Ibid.
10. Ibid.
11. Ibid.
12. Elise Boulding and Kenneth E. Boulding, *The Future: Images and Processes*, Thousand Oaks, CA: Sage, 1995, p. 100.
13. George Kurian and Graham T. T. Molitor, *Encyclopedia of the Future*, New York: Macmillan, 1996, p. 400.
14. Ibid.
15. Elise Boulding, *The Underside of History: A View of Women through Time*, Boulder, CO: Westview, 1976, p. 781.
16. Boulding and Boulding, *The Future: Images and Processes*.
17. Also *narancnici*, *orisnice* (Bulgarian) or *sudicki* (Czech). Spasoje Vasiljev, *Slovenska mitologija* (Slav mythology), Beograd: Velvet, 1996; Dusan Bandic, *Narodna Religija Srba u 100 pojmova* (100 Notions in Serbian folk religion), Beograd: Nolit, 1991.
18. Robert E. Bell, *Women of Classical Mythology*, New York: Oxford University Press, p. 310; Michael Grant and John Hazel, *Gods and Mortals in Classical Mythology*, Springfield, MA: G.& C. Merriam, 1973, p. 175. Owing to my 'broken' English I was surprised not to be able to find in these books any reference to ancient Nordic or Indian civilisation (I was not surprised there was no reference to the Slav tradition as our tradition rarely gets mentioned). Then I saw a book on non-classical mythology and thought, 'How interesting, what could contemporary mythology be?' My biggest surprise was that I saw references to classical and ancient Indian, Chinese and Nordic mythology—even a little bit on Slav mythology. Only then did I realise that only mythology from Greece and Rome deserves the name and the category of 'classical'.
19. Bell, *Women of Classical Mythology*.
20. Ibid.
21. Barbara G. Walker, *The Woman's Dictionary of Symbols and Sacred Objects*, San Francisco, CA: Harper & Row, 1988, p. 158.
22. Ibid. 460.
23. Ibid. 267.
24. Ibid. 266.
25. Ibid. 267.
26. Margaret Stutley and James Stutley, *Harper's Dictionary of Hinduism*, New York: Harper & Row, 1977, p. 137.

27. Walker, *The Woman's Dictionary of Symbols and Sacred Objects*, p. 16.
28. Ibid. 36.
29. Roy Willis, *World Mythology*, New York: Henry Holt and Company, 1993, p. 62.
30. Walker, *The Woman's Dictionary of Symbols and Sacred Objects*, p. 43.
31. Ibid.
32. One example is the previously mentioned *World Mythology* by Roy Willis. Although the author states that 'the goddesses of Egyptian mythology are often more formidable than the male deities' (p. 50) he does not allow them nearly as much space. He also dedicates a special section to 'powerful goddesses' (according to the tradition of the 'Women Question') only after many pages of description of male Gods.
33. *The Courier-Mail*, Brisbane, Saturday 21 February 1998, p. 29.
34. Ibid.
35. Patricia Aburdene and John Naisbitt, *Megatrends for Women*, New York: Villard Books, 1992, p. 262.
36. Ibid. 244.
37. Donna Haraway, 'A Cyborg Manifesto: Science, Technology, and Socialist-Feminism in the Late Twentieth Century', in *Simians, Cyborgs and Women: The Reinvention of Nature*, New York: Routledge, 1991, p. 181.
38. http://www.auburn.edu/~weissas/shrine
39. Ibid.
40. Ibid.
41. Boulding, *The Underside of History*; Elise Boulding, *Women: The Fifth World*, Foreign Policy Association, Headline series, 1980; Elise Boulding, *Building a Global Civil Culture: Education for an Interdependent World*, New York: Teachers College Press, 1988; Elise Boulding, *Women in the Twentieth Century World*, New York: Sage, 1977.
42. Riane Eisler, *The Chalice and the Blade: Our History, Our Future*, San Francisco, CA: HarperCollins, 1987; Riane Eisler, *Sacred Pleasure*, San Francisco, CA: HarperCollins, 1996; Riane Eisler, 'Dominator and Partnership Shifts', in Johan Galtung and Sohail Inayatullah (eds.), *Macrohistory and Macrohistorians*, Westport, CT and London: Praeger, 1997.
43. Francis Bartkowski, *Feminist Utopias*, Lincoln: University of Nebraska Press, 1989; Debra Halbert, 'Feminist Fabulation: Challenging the Boundaries of Fact and Fiction', in *The Manoa Journal of Fried and Half-Fried Ideas*, Honolulu: Hawaii Research Center for Futures Studies, 1994.
44. Judith A. Cook and Mary Margaret Fonow, 'Knowledge and Women's Interests: Issues of Epistemology and Methodology in Feminist Sociological Research', in Joyce McCarl Nielsen (ed.), *Feminist Research Methods*, Boulder, CO: Westview, 1990.
45. *Seminar*, 460: 13, December 1997.

6
De-Westernising Futures Studies

SUSANTHA GOONATILAKE

Futures studies has come in for a lot of criticism recently. The criticism concerns how futures studies deals or does not deal with issues and concerns of non-Western cultures and civilisations—the future of the bulk of the human population living outside the dominant political, economic and cultural regions of the world. These questions reveal both the academic inadequacy of the subject as well as the entrapment and disempowerment of the imagination of the non-Western victims of the futures studies industry. They also echo an exercise of a similar kind done nearly twenty years ago by Krishan Kumar, who analysed the futures studies of the time and demonstrated that in much of it, the non-Western world was marginalised and cognitively absent.[1] Twenty years later, Sardar has pointed out, in this volume and elsewhere, how the present global social structure of knowledge associated with the subject helps keep this partial cognition alive. My concern is with the following questions: how such a selective view of the future is socially constructed and, more importantly, maintained; and how one could step out of such restrictive formulations. In fact, what are the possibilities for, and restraints on the non-West freeing itself to do its own formulations?

The evidence that both Kumar and Sardar have marshalled is in the vein of the literature on the sociology of science, which has chartered in considerable detail how science in its creation, legitimation and application is suffused with a number of social and human concerns. Numerous studies on laboratory practice, journals, awarding of prizes and invisible colleges show science to be a very human enterprise, very much a *social* science. However, although this body of knowledge on the social construction of science has considered social forces at the level of the laboratory group, or even national political forces, it has not extended itself to the larger social forces of a geopolitical kind, or to a longer time-frame that takes into account historical factors. Thus, even the writings that have attempted to show the social dimensions of formal knowledge have limited their concerns to the Western world of the last few decades. Voices that have attempted to relate formal scientific knowledge to the non-Western world have been outsiders to discussions on science, as in the discussion on Orientalism,[2] or of the historical inadequacy of the usually assumed Greek-centred intellectual heritage.[3] And writers such as myself[4] who have discussed the geopolitical nature of science in the non-Western world have been largely peripheral to the Western debate on the social nature of science.

Yet, there is sufficient data to suggest how the geopolitical construction of knowledge takes place in the non-Western world, whether it be in the natural sciences or in the various versions of social sciences, including anthropology, development studies and, now, futures studies. The mechanisms by which this takes place are best described by considering the nature of scientific knowledge.[5] What is considered scientific knowledge in a dependent context is only what has been made legitimate in the centre. It is then imitated in the periphery through the operation of pervasive dependent social and cultural mechanisms. Generally, in such a model of knowledge fitting dependent countries, the process of knowledge-acquisition is largely a diffusionist one, even though the diffusion is sometimes limited only to the broad paradigms and not to minor variations within a paradigm. The fundamental and basic knowledge grows largely in the West and is transferred to developing countries in the context of a dependent intellectual relationship. The major paradigms, as well as the major problematics were and continue to be developed in the West (or at least legitimised there), and only minor variations of the major viewpoints are handled locally.

However, the fact that the Third World scientist has constraints that limit qualitatively his/her output does not imply a criticism at the individual level. The Third World scientist is hemmed in by a set of paradigms that were set elsewhere, with legitimation and reward systems that skew his/her work negatively and a total science system that in its operation is by its nature imitative and non-creative. However, Third World scientists transferred early to a First World centre quickly integrate themselves into the organic scientific community that they find themselves in and become productive, working on all knowledge frontiers.[6] Creative scholarship under such situations in the Third World is carried out under a siege mentality, battling odds on several fronts, with social and economic factors as well as data-availability restricting in a structured manner the 'freedom to imagine' of non-Westerners.

Conceptual boldness in such a dependent milieu will not be distinguished from a transplanted thought from the West that has recently arrived on the local intellectual scene and is also fresh to the local scene. Worse, lacking the package of legitimation that the latter brings, local boldness would not be seriously debated in the local academic literature. It might be captured by local sensationalist groups outside the formal legitimised knowledge groups and debated, but it would not enter the serious domain and hence influence formal discussions of legitimised Western subjects.

If these are the social mechanisms of formal knowledge, how could non-Western visions of the future emerge in its midst? What are the means of breaking through the restraints of legitimation structures that exist primarily in the West and the local social mechanisms of formal knowledge in developing countries that exist only to produce faded xerox copies of established positions elsewhere? This is especially difficult in such disciplines as anthropology, development studies and futures studies, which require a larger vision beyond those in natural science disciplines. In the latter cases, one could conceivably come across a single anomaly in a laboratory experiment that could upset established truths. Although even here, (social) psychologically to cognise such an event requires confidence to look beyond the structured legitimation channels.

Really to free oneself from the colonial bonds of a colonial discipline such as anthropology, literally a reversal of the subject may be required; in fact, a form of Europology has to emerge, a disciplinary mirror image of anthropology.[7] In the case of development studies, an ability to think beyond the ideological constraints of its constituent core disciplines, such as economics and sociology, is required, so that one can arrive at both alternative conceptual modes and alternative development paths. In the case of futures studies, which for the developing world has to be viewed as having the problems of development studies writ larger, still greater efforts are required. This implies that the Third World academics have to be highly sophisticated visionaries and that they have do more than their counterparts in the developed world. Because they have to have this larger synoptic vision, they have in many ways to work with a larger intellectual agenda in these disciplines—they have to become the non-Western equivalent of 'Renaissance persons'. But Renaissance persons in the non-Western world, unlike their European counterparts, have a doubly difficult task. They have not only to be aware of their society and its intellectual traditions, but also to operate in a field that is legitimised elsewhere—they must thus be aware of the West's intellectual environment too.[8]

One can identify two broad responses over the last century or so in the general interaction of non-Western and Western knowledge. One has been the occasional uses by Westerners of elements drawn from non-Western traditions as snippets and building-blocks for their own intellectual schemes. A recent example would include the elements drawn from non-Western sources that have fed psychology.[9] These acts one would call 'foraging'.[10] In contrast to foraging are legitimising acts by authors of the non-Western world who have attempted in such differing subjects as philosophy and psychology to show in the best Western universities that ancient non-Western searches sometimes coincided with current Western ones.[11] One useful strategy to make non-Western contributions to already established subjects would be for the legitimisers from the non-Western countries to become foragers for Western disciplines. They should ideally do this with the intellectual confidence that Western theorists have.[12]

Yet, even in the modern Western intellectual domain, a non-Westerner working in his/her own country is at a disadvantage. Usually s/he has no access to the current debates, and often the literature made available to him/her is dated. If it is a large country such as India—which has posed in a rigorous fashion some of the issues in the tension between the non-Western and Western intellectual traditions, as well as attempted seriously to develop the Western scientific tradition—there are possibilities of fresh thinking. India has a large number of knowledge practitioners, and so the possibility for pockets of non-imitative discourse is greater, although the search for new concepts in science in India's case too often results in 'aborted discoveries'.[13] But if it is a country such as Japan or China, where the tension between the Western and non-Western intellectual worlds was not seen as a possible two-way street with the non-West having potential uses in the quest for modernisation, then it is less possible to speculate freely. In such a situation, the Western intellectual discourse constantly hovers over one's shoulder as the only legitimate one, cramping one's style. The mind becomes closed to other, non-Western options, and only variants of the West emerge, in the forms of Chinese Marxist thinking or Japanese post-Meiji thinking (which in practice rejected the entire past Japanese intellectual tradition) or East Asian modernising theory. The probable consequence has been the relative lack of fresh thinking on the future in East Asia.

A small country such as Sri Lanka is more typical of the non-Western world. And its example illustrates the general difficulties of creative outcomes in the social situations I have sketched. Sri Lanka had debated some issues of the relationship between East and West for at least a century but is now, in spite of notable achievements, virtually a neo-colonial cultural entity. I will explore its example further, as it has many features in common with the bulk of developing countries.

Sri Lanka has near universal literacy, yet its flow of information at the tertiary levels and above, essential for informed debates, is probably less than in India. This is partly because, in several disciplines, university education has for more than a generation been in the national languages of Sinhalese and Tamil. This process has democratised access to education, yet it has restricted access to knowledge, largely because the literature in the local languages on current developments and disciplines is very limited and dated. In contrast, in India debates at the tertiary and post-tertiary levels are conducted after exposure to current Western debates, largely through the use of English-language literature; and in Japan and Korea a flood of translations keeps the local intelligentsia well informed about developments elsewhere. Furthermore, Sri Lankan science has had only luke-warm support from the state, unlike in India; and in the absence of a significant local bourgeoisie, by local benefactors. The combined result of these social factors is a highly skewed knowledge structure. In the late 1980s the university system had approximately 400 staff certified capable of independent social research; that is, with postgraduate degrees. But the total research money available annually for this social science community, a comparative study has shown, was just over $10,000! In contrast, a so-called non-governmental organisation—a private company—with a staff of only six with comparable backgrounds was receiving over fifty times the figure from foreign sources.[14] Consequently, essential resources in the form, say, of reading material and access to travel to meet other colleagues are denied to those with the best potential to use them in the highly literate Sri Lanka milieu. The result is that the collective social imagination of those who potentially would have the training to do social research is reined in.

The effect of these and other pervasive social mechanisms is that today Sri Lankan social thinking is very much governed, in spite of a few laudable local attempts, by external definitions. Thus, the Sri Lankan Studies Conference in Amsterdam in

1990 saw only a handful of Sri Lankan writers and nearly fifty Westerners, who between them have produced a relatively large literature. The latter pre-empts the 'international'—and hence, in a dependent situation, the local—academic definition of formal Sri Lankan social knowledge. This extends also to counter-definitions, which are faded mirror images of counter-positions in the West, including Marxism, modernity and postmodernism. The recent spate of writings on Sri Lanka, especially on the recent social and ethnic upheavals, is thus full not only of wrong interpretation but also of absolute untruths— much more so, one should add, than in colonial times. In fact, a recent book in this genre blatantly drew its title from a 19th-century hymn on the country that had the refrain 'Ceylon, where only man is vile'.[15]

If the general picture that I have drawn about the difficulties of practising the social imagination in non-Western contexts is so bleak, how can one really break through and give new formulations. Are there possibilities? Are there examples?

The need to incorporate extra-European cultural elements also arises from the requirement to replenish what has been described by many as the exhausted Enlightenment project of Western civilisation in several core intellectual areas and hence to give a meaningful set of new rudders for the future. This exhaustion is seen in the questioning of the seeming epistemological and ontological certainty that came during the Scientific Revolution and gelled into the Enlightenment. First is a quantitative challenge.

Science's output has doubled every ten years or so since the mid-17th century, and will soon be far beyond the capacity of humans.[16] Already projects like the Hubble Telescope and the Human Genome Project are spewing out so much raw data that it has been compared to 'drinking from a fire hose'.[17] The solution being sought is extensive automation in the laboratory, data-collection and analysis, including through the use of 'Knowbots' and 'Discovery Machines'—now being researched—that scour data terrains to emerge with conclusions.[18] Already programs have rediscovered scientific laws like those of Ohm and Snell. So, it seems that the Enlightenment dream of Diderot to capture all the available knowledge in the West is, because of databases, on the verge of realisation. Given this extent of automation, the agenda of the 17th-century Scientific Revolution can be left partly to machines. This again begs the need to enlarge the epistemological project.

The epistemological world of molecular biologists is not the same as that of evolutionists or of taxonomists. The Newtonian mechanics used in most astronomy and mechanical engineering assumes the reversibility of time. On the other hand, thermodynamics has a time direction. Those who deal with quantum and relativistic phenomena are in a logical, epistemological and ontological world different from that decreed by the ancient Greeks. 'C + C' in the relativistic world is still C, the speed of light; and in the quantum world a particle both is and is not, just as it is a wave and is not. The key ontological Cartesian dichotomy of subject and object has come unstuck in quantum physics. The post-Galileo project of true knowledge through mathematisation has come into question, because, ever since

Gödel, mathematics has lost its foundational certainty. And Bacon's vision of science as torturing nature to reveal its secrets has run its course, because tortured nature's environment is under severe strain.

So, a variety of cultural, epistemological and even ontological positions jostle in today's knowledge system. This is seen most vividly in the social sciences. Social sciences rose after the Enlightenment and were ideological adjuncts to the rise of Europe as a hegemonising system. Anthropology, Sinology and Indology, which saw non-Europeans as backward, reflected this Eurocentrism. For example, Marx's residual category the Asiatic mode of production and Weber's equating Protestantism with the rise of capitalism also saw Asia as backward. Yet, today's Asiatic modes of production are the world's economic dynamos (the current setbacks are essentially temporary), which requires an urgent rethinking for the social sciences.

The ultimate intellectual culmination of the Enlightenment project was in many ways modernism. Its demise and the emergence of postmodernism are a reflection of the exhaustion of the Enlightenment project on the terms and agendas set by itself. So extra-Enlightenment elements become imperative for navigating the future. There are already signs. With the centre of gravity of the global economy shifting to Asia within the next twenty years, a geotectonic shift comparable to the 'Voyages of Discovery', the Renaissance, the Scientific Revolution, the Industrial Revolution and the Enlightenment all rolled into one and compressed into a couple of decades is occurring. This shift will, without question, ease and in fact necessitate the use of non-Western intellectual fodder.

To begin with there are holes in the hegemonic blanket. Postmodernism and deconstruction, with their penchant for casting doubt on established truths, are a fact of the current Western intellectual terrain. They are, however, for the non-West both symptoms (loss of certainty) and an opportunity (delegitimation of existing truths and hence possible openings for other intellectual interventions). And this potential openness allows fresh conceptual elements from other cultures to be introduced, as long as they pass the test of usefulness. The world has already become partially sensitised to the values of non-Western knowledge—as, say, in the knowledge of plants held by the inhabitants of the rainforest, which is valuable for the biotechnology enterprise. Although there is partial legitimation of the usefulness of knowledge of such small groups at the hunter–gatherer stage, paradoxically, the potential contributions of the more formalised knowledge of some of the larger non-Western civilisations have not yet been adequately recognised.

There are, indeed, possible areas of contribution by such non-Western-civilisational stores of knowledge to the contemporary scientific knowledge enterprise that could enlarge current intellectual horizons, including those of visions of the future. These include intellectual areas that would have been explored in other traditions but not in the Western tradition. These explorations are cognitively lost to modern science just as several times in the Western tradition too there have been similar cognitive losses. These one-time losses in the Western context have varied from atomism, to the dust-cloud hypothesis of the solar system's formation, to

Mendelian genetics, all of which were cognitively lost once but were later resurrected. In a similar fashion, there are areas of exploration in other traditions that are today cognitively lost. Hence the potential importance of, say, the Islamic scientific tradition today is not in patently spurious attempts to explain its theology in a 'scientific' manner (as have been made by some who identify themselves with the Islamic science movement), but in the fact that there may be in some of the Islamic manuscripts of an earlier date areas that were explored in mathematics or medicine, for example, that could have contemporary uses.

To illustrate this possibility, let me give some examples. In the conventional historiography of science, a few figures stand out as key turning-points. Among them are William Harvey, who described the circulation of blood, Paracelsus, who epitomised chemistry in the Renaissance and developed a salt-based alchemy, Copernicus, who described a heliocentric system, and Linnaeus, who established a classification system in biology. Now, recent research in non-Western cultures indicates that all these achievements were in various forms known to other cultures. The European genius was either in an independent discovery or in the assemblage of others' discoveries under one intellectual umbrella, which the 'Voyages of Discovery' and the new global gaze allowed Europe to do.

Let me elaborate. The circulation of blood was known to the Islamic tradition at least centuries before Harvey; the Muslim biologist ibn Nafis first described the circulation of blood through the body in the late 13th century.[19] A salt-based alchemy was being taught in Saudi Arabia in the 12th century by a South Asian expatriate.[20] It has been shown that the astronomers at Maraga had developed heliocentric mathematics—including the Tusi couple attributed to Nasiruddin al-Tusi—in the 14th century, and a recently discovered 15th-century Kerala manuscript has evidence that a heliocentric mathematics was being used there at least half a century before Copernicus.[21] Work by ethnobiologists suggests that the Linnean classification system had less to do with Linnaeus' own individual genius in classification than with the fact that he had a larger set of biological samples. Research has shown that hunter–gatherer groups and modern Linnean taxonomists arrive at near identical classifications when sampling the same forest area.[22] If these examples, which show that at least partial parallels of apparent key turning-points in the modern knowledge enterprise existed previously elsewhere, are any guide, then there could be potential contributions from other intellectual traditions. This suggests the possibility of actively exploring this cognitively hidden knowledge.

There have been several recent attempts by non-Western traditions to influence debates in Western science. One recent attempt is by the biologist-cum-cognitive scientist Francisco Varela, who has used insights from Buddhist meditational practice on the 'embodied mind' to suggest solutions to the mind–body problem in the cognitive sciences and the organism–environment issues in evolutionary theory.[23] In fact, based on precisely such cross-civilisational assumptions, there is a major project in India, sponsored by its regular scientific institutions, to explore the past traditions for material useful for contemporary knowledge, including algorithms for artificial intelligence.[24] There are numerous other examples from a host of disciplines—the non-Western-civilisational stores are pregnant with countless possibilities.

Elsewhere,[25] I have explored these possibilities in some detail. Here, I give a few examples from medicine, mathematics, physics and psychology. The classical methodology of the Ayurvedic medical systems has several knowledge-acquisition techniques for coming to verifiable conclusions from known facts. Modern Western medicine is using these techniques, and current research in the Western medical tradition is legitimising several Ayurvedic practices in the treatment of, among other things, wounds and ulcers, diarrhoea and stomach diseases, liver ailments, the reproductive system and contraception, the nervous system, cancers, the heart, the lungs and the cardio-vascular system. There have also been estimates of the reservoir of Ayurvedic knowledge yet unexplored that indicate a very large potential for therapeutic uses. International organisations such the World Health Organization have made preliminary forays into listing further possibilities of potential cures in the field.

In mathematics, we know that there have been several different approaches, methodologies and traditions—such as Greek, Chinese, Indian and Arab—which have cross-fertilised each other in the past. In the case of South Asia, there is ample evidence to suggest that after the Arab connection to the West was completed new approaches in mathematics continued to flourish. These included aspects of permutations and combinations, indeterminate analysis, infinity, the binomial theorem, impetus theory, gravitation, Newton's first law, calculus, set theory and heliocentric astronomy. Contemporary searches in South Asian traditions have included developing arithmetic beyond its classical South Asian imprint as taught all over the world today, extending some of the results of the old algebraist Bhaskara, programming a computer-based language-translation system using the theoretical insights of Pannini (a 5th-century BC grammarian), and deriving certain characteristics of quantum theory using Buddhist philosophical theories. Estimates of the hidden mathematical knowledge to be found in South Asian manuscripts indicate that these efforts have only just scratched the surface.

Among the most important of recent South Asian influences have been those in psychology and the development of mind–body medicine. With the collapse of the speculative and partial approaches of Freud, Jung and Skinner, South Asian inputs tested on rigid scientific criteria have begun to fill the gap. These additions include the variety of South Asian mental techniques known as 'meditation', which have been studied in great detail, including by the use of modern brain-imaging techniques. The approaches that have been found to have considerable efficacy include mindfulness meditation, Transcendental Meditation and various imagery techniques. They have been applied clinically in behavioural change, stress treatment, anxiety, panic and phobias. These and other South Asian approaches have been increasingly legitimised in both theory and practice and these results are found today in many of the standard technical journals dealing with psychology and medicine.

The emerging technologies of biotechnology and information technology raise important questions of human identity and challenge conventional notions of the self, which becomes problematic with their applications. This has raised a host of ethical and philosophical questions in areas such as the bioethics of transplants, reproductive technology, implants, prosthetics and cyborgs. Answers to these questions have direct relevance to, and possible solutions in the vast Buddhist and Sufi philosophical literature on the self. Virtual reality, too, raises important questions on the nature of the 'reality' that it generates and its relation to our everyday reality, as has been noted by the more philosophical writers on the field. Once again, the questions raised here have distinct echoes in some of the discussions on reality in classical South Asian philosophical systems such as the Upanishads, Jainism, Buddhism, Nyaya Vaisheshika and Vedanta. I would suggest that inputs from these South Asian explorations would be of immense benefit to emerging issues in virtual reality.

The South Asian region has, in its different schools of philosophy, explored a tapestry of problems different from those in the Western philosophical trajectory. There are new philosophical questions coming up in several scientific fields. These include the questions of logic and causality in quantum physics, varieties of systems thinking, debates on general evolution, including questions raised in the field of bio-epistemology, cognitive science and artificial intelligence. There is a great potential for South Asian inputs into these debates—for example the potential of Buddhist fourfold logic or Jain sevenfold logic to solve the new logical problems in physics, or of Buddhist process philosophy in evolution. The inevitable conclusion seems to be that there is a vast reservoir of non-Western knowledge to be tapped. Estimates indicate that only a very small number of available old manuscripts have even been translated from the Sanskrit, let alone accessed by modern scientists. The future seems to be promising and exciting.

The examples I have given of the structural reasons for the non-Western voice not being heard, and of how creativity in the modern knowledge enterprise could be enlarged by incorporating local knowledge traditions, have a direct relevance to futures studies. They allow the essential cultural elements from within the non-Western traditions themselves to interact with the Western-oriented one, thus enlarging and enriching the field. These elements include both the local historical experiences and the conceptual elements that have emerged out of that experience. Only by such means can the non-Westerner enter into meaningful and non-imitative discourses about the future. There are examples of this, in subjects related to futures studies where disciplines are trying to break out of the strait-jackets, in the emergent writings of some individuals exposed to the workings of both the non-Western and Western intellectual projects. Especially in some regions of South Asia and the Islamic countries, there are several new voices now emerging. Current technological and social tendencies towards globalisation potentially allow such individuals to carve out interacting groups, outside the existing structured ones. These individuals are beginning to transcend both the entrapment of the existing geopolitically biased knowledge and the ossification of their own cultures. They could thus potentially traverse with freshness the new globalised world's intellectual niches in a way not possible earlier. Existing studies within the sociology of knowledge of how once-marginal groups have become legitimate, as for example by having their own rigorous journals, could be guides for the future development of these groups.

The lineage of futures studies can be traced back to the confident utopian constructions of the Renaissance and after and the science fiction narratives of 19th-century Europe. It purports today not to have a normative element and wishes to be considered objective and scientific. But in capturing the fate of the future, it also captures 'today's' (meaning in actuality the West's) values, prejudices and presumptions. It is reality and hope, map and goal. For non-Westerners, the imperative to capture the high ground of imagination is both to free themselves from the intellectual holds of others and to realise themselves socially.

Notes

1. Krishan Kumar, *Prophesy and Progress: The Sociology of Industrial and Post-Industrial Society*, Harmondsworth: Penguin Books, 1978.
2. Edward Said, *Orientalism*, New York: Random House, 1978.
3. Martin Black Bernal, *Athena: The Afro-Asiatic Roots of Classical Civilization*, London: Free Press, 1987.
4. Susantha Goonatilake, *Aborted Discovery: Science and Creativity in the Third World*, London: Zed, 1984, pp. 157–68.
5. This discussion is taken from Susantha Goonatilake, 'Development Thinking as Cultural Neo-Colonialism', *Bulletin of the Institute of Development Studies*, April 1975.
6. Susantha Goonatilake, 'Epistemology and Ideology in Science, Technology and Development', in Atul Wad (ed.), *Science, Technology and Development*, Boulder, CO: Westview Press, 1988, pp. 93–115.
7. Susantha Goonatilake, 'Towards a Study of Europology', Presidential Address, Social Sciences Section, *Proceedings of the Sri Lanka Association for the Advancement of Science Annual Sessions*, December 1976.
8. Susantha Goonatilake, *Crippled Minds: An Exploration into Colonial Culture*, New Delhi: Vikas, 1982.
9. Susantha Goonatilake, 'The Voyages of Discovery and the Loss and Re-Discovery of "Other's" Knowledge', *Impact of Science on Society*, Paris and London: UNESCO and Taylor & Francis, 42(3): 241–64; Dale Riepe, 'The Indian Influence in American Philosophy: Emerson to Moore', *Philosophy East and West*, 17(14): 124–37, January–October 1967; A. H. Maslow, *Towards a Psychology of Being*, New York: Van Nostrad, 1968, p. 141.
10. Goonatilake, *Crippled Minds*, pp. 265–78.
11. Ibid. 278–90.
12. Goonatilake, *Aborted Discovery*, pp. 162–3.

13. Ibid. 91–119.
14. Susantha Goonatilake, 'The Science System in Sri Lanka', in A. Rahman (ed.), *Science and Technology in the Indian Subcontinent*, London: Longman, 1990.
15. William McGowan, *Only Man is Vile: The Tragedy of Sri Lanka*, New York: Farrar, Straus & Giroux, 1992.
16. D. J. de Solla Price, *Little Science, Big Science*, London: Macmillan, 1963; Science Foundation Course Team, *Science and Society*, Science Foundation Course units 33–4, Milton Keynes: Open University, 1971.
17. M. Mitchell Waldrop, 'Learning to Drink from a Fire Hose', *Science*, 248: 674–5, 11 May 1990.
18. Ibid.
19. Seyyed Nasr, *Science and Civilization in Islam*, Cambridge, MA: Harvard University Press, 1968, p. 154.
20. D. M. Bose, S. N. Sen and B. V. Subarayappa, *A Concise History of Science in India*, New Delhi: Indian National Science Academy, 1971, pp. 334–8.
21. K. Ramasubramanium, M. D. Sirinivas and M. S. Sriram, 'Modification of the Earlier Indian Planetary Theory by the Kerala Astronomers (*c.*1500 AD) and the Implied Heliocentric Picture of Planetary Motion', *Current Science*, 66(10): 784, 25 May 1994.
22. Brent Berlin, 'The Chicken and the Egg-Head Revisited: Further Evidence for the Intellectualist Bases of Ethnobiological Classification', in Darrel A. Posey and William Leslie Overal (eds.), *Ethnobiology: Implications and Applications. Proceedings of the First International Congress of Ethnobiology*, Belem, Brazil: Museno Emilio Goeldi, 1990, pp. 19–35.
23. Francisco J. Varela, Evan Thompson and Eleanor Rosch, *The Embodied Mind: Cognitive Science and Human Experience*, Cambridge, MA: MIT Press, 1991.
24. Navjyoti Singh, *Research Proposal Foundations and Methodology of Theoretical Sciences*, New Delhi: NISTADS, 1990.
25. Susantha Goonatilake, *Toward a Global Science: Mining Civilizational Knowledge*, Indiana University Press, 1998; Susantha Goonatilake, *Merged Evolution: The Long Term Implications of Information Technology and Biotechnology*, London: Gordon & Breach, 1998.

7
Implementing Critical Futures Studies

RICHARD A. SLAUGHTER

Critical futures studies is steadily emerging from its somewhat esoteric origins, but it has yet to become mainstream. Given the unstable conditions of the late 20th century, and the challenging outlook of the early 21st, this is a significant oversight. However, this chapter does not take the usual approach of beginning from an outline of the nature and benefits of critical futures work. Rather, it takes a more structural approach and considers how such studies can be progressively developed through four distinct layers, or levels. First is the natural capacity of the human brain–mind system to envisage a range of futures. Second is the clarifying, enlivening and motivating role of futures concepts and ideas. Third are analytic gains provided by futures tools and methods. Fourth are a range of practical and intellectual applications, or contexts. When all of these levels function in a co-ordinated way, grounds for the emergence of foresight at the social level can clearly be seen.

At first sight the future is a highly problematic field of study that is unrelated to the problems and concerns of 'ordinary' people. How, it is asked, may one study something that doesn't exist? Academic futurists respond to this basic challenge in various ways. For example, they may point out that futures studies deals with intangible phenomena—as do aesthetics, law, ethics and religion. Others suggest that futures studies is essentially about how present-day ideas, feelings, goals etc. influence the future. Still others focus on the creation of 'surrogate', or 'interpretative' knowledge about the future that takes the place of future facts. In this latter view the future can be said to exist—not as an empirical, measurable realm, but as one of vision, will, understanding and interpretation. Hence futures studies is more closely related to the social sciences (and vice versa) than to the so-called hard sciences. It is therefore reasonable to think of the future as 'a principle of present action', because this highlights the way past, present and future interact.

However, for most people academic approaches to the 'the future' mean that it remains an abstraction. While stereotypical images of futures are widely available in popular culture, few people take them seriously or consider the much wider range of images and social trajectories that are available. Equally, the rich links between values, paradigms, ways of knowing and the future are overlooked—even by mainstream futurists. Thus, for the majority, the future might as well be 'an empty space' for all the effect it has on their daily lives and decisions, their personal and professional behaviour.

It is for such reasons that more robust and penetrating approaches to futures work are needed: approaches that are intensely practical on the one hand and, on the other, capable of dealing with issues in depth, reflexively aware of embedded cultural commitments and also committed to the decolonisation of futures work—the move away from unquestioned cultural and ideological interests to more universal ones. To be useful, futures work must be accessible. To be accessible it must have meaning for large numbers of people. Hence a structural approach may be more useful than an academic one.

At present a largely American, empiricist and ideologically shallow model of futures work prevails, especially in corporate and business contexts. Hence governments around the world still maintain their short-term time-horizons up to the next election, with little or no thought for the longer-term implications of the major shifts under way, the period of fundamental transition collectively facing us in the 21st century and the plight of future generations. How, then, could this apparent abstraction, 'the future', be made more real, more accessible, more a part of daily life? How can a technology-obsessed culture rein in its dynamism and listen to the more subtle voices of the natural world and the needs of future generations? How can futures studies be decolonised and a range of new and original voices be brought into the international futures discourse? I do not think that such outcomes can be achieved by threats, gloom-and-doom posturing or any expectation that governance will be transformed in the near future. A different strategy is needed, one that recognises the layered quality of futures understanding. This cannot be legislated into existence. However, this chapter explores the view that it can be built up layer by layer over a period of time.

Level 1: Human Capacities and Perceptions

A sound place to begin is with individual human capacities. It is evident that the human brain–mind system is richly endowed with the capacity not just for primary consciousness (seeing only what is directly available to the senses) but for reflexive understanding in time. This higher-order consciousness is characterised by the ability to remember and to learn, to roam consciously throughout a rich, complex, extended present, to understand responsibilities and consequences, and to speculate on futures yet to come. Edelman characterises it this way:

> The freeing of parts of conscious thought from the constraints of an immediate present and the increased richness of social communication allow for the anticipation of future states and for planned behaviour. With that ability come the abilities to model the world, to make explicit comparisons and to weigh outcomes; through such comparisons comes the possibility of reorganising plans. Obviously, these capabilities have adaptive value.[1]

Human beings therefore have an innate capacity for speculation, foresight, modelling and choosing between alternatives. They are not stranded, willy-nilly, in a deterministic world. Rather, they are consciously located in a socially created, but self-actualised matrix of structures, understandings and forces. It is for such reasons that human beings are able to think about not only 'the future' but 'futures' plural. Unlike the human body, which is necessarily constrained in time by the close co-ordination of biology (respiration, digestion, protein synthesis), the human mind, imagination and spirit are free to roam at will among a stunning array of different worlds and worldviews, past, present and future. They can also communicate directly with future generations.

Crudely put, the 'wiring' of the brain–mind system is sufficiently complex and inclusive to permit at least three kinds of journey. It routinely permits consideration of past environments that the body and perceptual apparatus were never present in to experience directly; it supports knowledge and understanding of significant contexts in the historical present that are displaced in space (e.g. Chernobyl, Bosnia, Oklahoma City); and it enables the forward view—a potentially panoramic outlook on a vast span of alternative futures. Therefore, capacities and perceptions are two of the 'building-blocks' of futures studies in general and the future-generations perspective in particular. The ability to think ahead is grounded in these features. It is an emergent capacity of this complex, elegant system. This is why all normal persons are fundamentally capable of foresight, forward thinking and responsible behaviour focused on long-term considerations. In contrast to the professional (and professionalised) work of, say, forecasters and scenario-analysts, one does not need a Ph.D. and an academic base in order to engage in long-term thinking.

Level 2: Futures Concepts Enable a Futures Discourse

However, the raw capacity of the brain–mind system clearly does not automatically lead to understanding and competence. The rare stories of children raised by animals give weight to the view that to become properly human, the young need to be nurtured within a family and inducted into the symbolic social world of language and culture. So far, so good. Unfortunately, however, those raised in Western, or Westernised cultures, are likely to be imbued with the characteristic Western outlook: nature is purely utilitarian—it is merely a resource for human use; growth is seen as an unproblematic and unquestioned good; science and technology are primary forces in creating 'opportunity' and hence the future; the cultural past is valued and tangible but the future is not similarly regarded.[2]

Critical futures studies makes it clear that such embedded cultural commitments are complicit in the emergence of the global problematique in all its many dimensions. Therefore, each generation that takes on such commitments and assumptions, that regards them as natural and normal, perpetuates an unsustainable world order at a very fundamental level. However, as noted, higher-order consciousness is reflexive. It can look clearly on its own presuppositions and, where the evidence is clear, change them. This work is obviously not easy, but it is certainly possible over a period of time. As this occurs, so the 'mist' clears, and a diagnosis can emerge about the global plight of humankind. Unfortunately, this is as far as many scholars and others get. Yet the next step is as simple as it is powerful: the development of a personal futures discourse.

It is tempting to see such a discourse as simply a matter of acquiring the appropriate language. There is some truth in this, but it is not the whole picture. As a teenager I read a great deal of science fiction. Without realising it at the time, I was learning a grammar of futures imagery that subverted the usual default view of the future as a blank space and instead populated it with an immense variety of images, meanings and possibilities. Because of this variety, my relationship to the future became active, rather than passive. For example, I began to wonder why so many fictional futures were populated by cruel aliens, implacably menacing supercomputers, rampaging robots and earthly catastrophes. Why, I wondered, could the future not be a desirable place, even in imagination? The answers to such questions led me to the futures field, its rich, inspiring literature and, eventually, to some of the people who created both.

When I later began to explore the world of futures studies, I came to see that there was indeed a distinct discourse. As I began to immerse myself in it, so many aspects of the futures dimension began to clarify and to connect with features of the present that implied particular directions, outcomes and scenarios. This provided a new purchase on current affairs. Eventually I could see that society is profoundly affected by a small number of dominant discourses that, in no small way, condition the framing of current issues and concerns—and hence the priorities and directions adopted at any time. One of the dominant discourses is an economic one. It heavily influences how governments govern, allocate resources and make decisions. However, it is predicated on a range of untenable assumptions that have been thoroughly critiqued by futurists and others.[3] Another dominant discourse, not unconnected with the above, is the commercial one. This is based on equally untenable assumptions. It essentially says, 'Buy, consume, use and use up everything you want. Give no thought for tomorrow.' A third discourse is academic. It is deeply conservative and committed to boundary-maintenance. Here future-discounting is very strong. Academia values the past much, much more deeply than the future. One could go on, but I think the point is made. Set against these are a number of newer discourses that are engaged in a symbolic struggle for acceptance —for example, a peace discourse, an environmental discourse, a discourse arising from the women's movement and a multicultural decolonisation discourse. Each of these attempts to legitimise particular concerns through language.

The futures discourse shares in the need to achieve acceptance and legitimation. But it is less clearly focused on achieving specific cultural goals. The most specific generalisations that could, perhaps, be derived are the need for a shift from short- to long-term thinking and the notion of sustainability as a social goal.[4] However, beyond this it does not appear to be strongly prescriptive. One reason may be that the core concept of 'alternatives' mitigates against such an approach. Yet I think it true to say that the lack of a futures discourse in society is one of the structural impediments to adaptive change. Or, put positively, the wider uptake of the discourse is one of the most powerful strategies for dealing with the apparently intractable dilemmas of the present and near-term future.

Without a futures discourse founded on the critical thinking alluded to above, 'the future' is occluded, hidden, continually just out of sight and therefore out of mind. People therefore just don't think about it. I call this the 'threshold problem'. The many rich possibilities for understanding the global predicament, reconceptualising aspects of it and steering towards consciously chosen outcomes are therefore overlooked. So what can be done? A number of futures scholars have tried to make futures concepts and ideas more widely available in the belief that, in the process, the social capacity to use and apply the discourse will be enhanced.[5] What evidence is there for this view? It is twofold. First, there is the personal experience of a dawning awareness followed much later by a progressively deeper understanding. Second, there is the experience of graduate students from around the world who came fresh to futures studies and finished their courses with a much more empowered and insightful view. I do not want to underestimate the problems that students new to futures studies may experience, but I have no doubt whatever of the enabling power of a futures discourse.[6]

The most broadly useful futures concepts are those that have a certain 'amplitude'; that is, they can be approached and understood on a variety of levels. They can therefore be introduced to young children as well as adults. For example, the notion of foresight may seem difficult. But it can be approached concretely by observing its use in everyday life (walking, driving, sailing etc.). Later, the notion of a 'loop of futures scanning' can be introduced. Still later aspects of systems thinking and theories of human perception come into view. For these and other reasons, foresight is one of the most productive futures concepts I have ever encountered (which is why I wrote a book about it). Some others are:

- alternatives and choices
- breakdown and renewal
- cultural editing
- empowerment
- extended present
- futures in education
- future generations
- reflexivity
- social innovations
- sustainability
- time-frames

- vision
- a wise culture

It might be objected that not all such concepts are 'owned' by futurists. Quite so. But when they are used in a sustained way and in combination with others, as well as with the other resources available through futures studies, they do, I believe, permit a distinctly futures-oriented quality of understanding to emerge. It is this that is the goal and purpose of critical futures educators, rather than the pursuit of this or that particular future scenario. Thus, futures concepts enable a futures discourse. It is the latter that provides the foundation for an applied futures perspective, rather than techniques per se. However, as the following section suggests, methodologies do have an essential role to play.

Level 3: Futures Methodologies and Tools

If it is innate capacities that make futures thinking possible and futures concepts that enable a futures discourse to emerge, it is the use of tools and methodologies that raises the power of a futures perspective to a new level. It is all very well to articulate futures issues and problems. But at the end of the day discourse alone cannot deal adequately with many broader or more complex futures concerns. For example, take the practical need to assess if a power station or a new major road is, or is not, necessary. Making a sound decision depends not merely on ideas and discourse, but also on the extended treatment of complex sets of data. This is where a purely literary-based discourse reaches its limits: it cannot handle data. But some futures methodologies have been developed for precisely this purpose, to generate, manipulate and evaluate concepts, ideas and information about the future. They include:

- backcasting
- cross-impact matrices
- Delphi surveys
- environmental scanning
- forecasting
- the futurescan process
- layered causal analysis
- scenario-building
- trend-analysis
- visioning

In the case of building a major road or airport, much work would be needed to collect time-series data from the recent past and then subject it to various forms of mathematical extension. It is here that scenarios can also be used to embed the raw data in humanly meaningful contexts, that is, self-consistent pictures of possible futures. From this kind of elaboration emerges a view of the context in which the proposed project can be located. Does it still make sense? Is it 'economic'? Are the trends likely to hold up? What 'system breaks' can be envisaged? And so on. It is clearly a very demanding and sophisticated exercise that requires expert knowledge and understanding. Despite some corporate overtones, the work of Peter Schwartz and Jay Ogilvy at GBN has carried this kind of approach to a high level of accomplishment.[7] It is both commercially and intellectually successful, providing many organisations with valuable strategic intelligence about their business, products and markets. Properly handled, scenario-planning can, perhaps, also assist in the transformation of these industrial-era entities. In other words, the manipulation of large data-sets in combination with conceptual sophistication allows a kind of extended analysis to take place that deals successfully with complex practical problems.

There is also another aspect to this level: futures tools. By tools I refer to simpler strategies and procedures that serve to extend our understanding in a wide range of situation, including:

- assessing global 'health'
- brainstorming
- the critique of images of futures
- dealing with young people's fears
- exploring the extended present
- futures wheels
- imaging workshops
- the loop of futures scanning
- questions about futures
- simple cross-impact matrices
- simple scenarios
- simple technology-assessment
- simple trend-analysis
- social innovations process
- time capsules
- time-lines
- values-clarification

Such futures tools have been derived from the conceptual and methodological resources of the futures field by many people: educators, social activists, workshop facilitators and others. Metaphorically speaking, they provide a very comprehensive 'tool-kit' for those working with young people or in other enabling contexts. Various attempts have been made to make some of the most useful tools available in published form.[8]

Even taken alone such tools can be extremely useful. Take the futures wheel. It is a very simple idea. A possible future event is placed in the centre of a piece of paper. Immediate consequences are traced out in a rough circle. These 'first-order' items are then explored in another ring and so on. Practitioners have found it an ideal starting-point to investigate the implications of many topics with people of all ages, from very young children to corporate executives. In other words, it has the 'amplitude' I mentioned above. It can also be used as a mind-map, a counselling tool and a way of exploring assumptions.

But futures tools do not exist in isolation. They can be assembled in many varied and productive sequences. This is partly why they are such a useful and flexible educational resource. For example, one can begin with an exercise dealing with optimism and pessimism, go on to one dealing with young people's fears, continue with one on social innovations and end with another using simple scenarios. From this basis, a second sequence could examine some more advanced concepts and ideas: the critique of the industrial worldview, the nature of a wise culture, communicating with and caring about future generations. The subjects and permutations are endless. Clearly such tools involve and complement the use of futures concepts.

Thus futures methodologies and tools greatly enhance the development of a futures perspective by extending the analytic, cognitive and intellectual reach of those using them.

Level 4: Futures Applications

All of the above would be very limited in scope if such resources were only used in an ad hoc, informal way. But they take on much greater force and power when embodied in specific contextual applications, for example:

- critical futures studies
- future-generations studies
- futures in education
- futures research organisations
- institutions of foresight
- strategic planning units
- 21st-century studies
- university futures departments

There is some overlap here with the implementation of the more sophisticated methodologies, since the latter obviously require a context. But it is the term 'context' that is significant. It does not matter how articulate we may become, or how far-reaching our methodologies may be. If there is no supportive context, such powers will be difficult to sustain; they may wither and die. This is the fate of all-too-many futures-related educational innovations. So the key to this level is the provision of an institutional or organisational milieu where high-level futures work can thrive, develop and be critiqued and implemented. I will now briefly discuss the examples given.

Critical futures studies uses the standard tools of scholarship to raise the power of futures thinking to a higher level.[9] However, it cannot take place in isolation. There is no point in gaining insight if the insights are private, unavailable and not susceptible to the very necessary process of disciplinary criticism and feedback. For me, one key to implementing some of the insights from critical futures study was to embed them in postgraduate university courses.[10] This had the benefit of testing them against the criteria of tertiary institutions and also against the needs and perceptions of successive cohorts of postgraduate students. It is essential that such work is openly tested and does not become a private indulgence. However, critical futures work is not merely an academic approach; it is intensely practical as well.

Future-generations studies is a new field that has emerged from futures studies per se and from the application of particular values within a futures perspective. In this view, future generations are radically disadvantaged by many present trends, structures and practices. Therefore, ways are needed to raise the profile of future generations, to give them a voice, and for contemporary societies to recognise their needs in present-day councils. This necessary social innovation is given even more weight when it is realised that many traditional cultures in the past already understood the necessity of so doing.[11]

More generally, much of the foregoing can be implemented in education at the school and college level. I have always argued that instead of being seen as a brash newcomer, a futures perspective is intrinsic to the tasks of teaching, learning, teacher preparation and professional development (particularly for principals). Schools are in the business of preparing the next generation for life in the early 21st century. They are therefore one of the few social institutions with a social mandate to think long term, though few of them yet know how to do it. Still, there are signs of progress. As this chapter is completed, the Board of Senior Secondary School Studies in Queensland, Australia, has completed the first trial of a new, four-semester futures syllabus for year 11 and 12 students. Such innovations will multiply during the millennium period. The opportunities for productive work in this area are much greater than is commonly recognised.[12]

Futures research organisations are purposely designed to facilitate the kind of extended, high-level, data-driven work noted above. They have sprung up in many countries, particularly in the USA, where a combination of wealth, entrepreneurial drive and deep-seated perception of opportunities and problems has created a critical mass of expertise. Such institutes tend, on the whole, to carry out work for government departments, public utilities and corporate clients. What I call 'institutions of foresight' (IOFs) are closely related to them. They may use some of the same methods. However, their values tend to be different and the focus is more in the public interest arena. Depending on how they are defined, there are several hundred IOFs around the world.[13] They exist because perceptive people in all regions have understood the drift of world events, have seen the outlines of the 'great transition' that lies ahead, and have realised that no country should simply drift into this most dangerous and challenging time. This discovery gives weight to what I think of as a growing 'congruence of insight' about the fundamental problems facing humanity, as well as long-term, systemic solutions. Hence the research institutions and the IOFs play a potentially central role in helping to embed society-wide changes of perception and practice.

Another milieu in which such work may be done is the strategic-planning units of various organisations. While strategic planning has not always lived up to the expectations loaded upon it, most large organisations have discovered that they must try to operate strategically. Those who do not are much more likely to fail. So strategic planning is not likely to be abandoned. It can, however, be improved by opening it up to the kinds of symbolic and methodological resources outlined here. Too often, for example, it is the case that corporate approaches to futures are epistemologically and ideologically naive, taking, say, a particular corporate or cultural ideology as 'given' and missing altogether the many options for critical analysis and reconceptualisation upon which lasting social innovations may depend. Some management books fall into exactly this trap, including Hamil and Prahalad's *Competing for the Future*. Others, such as Paul Hawken's *The Ecology of Commerce*, deal with the coming transition in much more conceptually adroit ways.[14]

The emergence of 21st-century studies is potentially one of the most significant developments in the field, though it is not yet widely known. A number of national studies have been carried out. A conceptual and methodological tool-kit has been developed, along with an enviable grasp of the organisational and practical issues that can determine success or failure.[15] It is regrettable that only a small number of nations have so far participated in this programme. Many rich Western nations, including Australia, continue to proceed in blissful ignorance of the high-quality international 'conversation' that they are missing. In time, the national studies will form the basis of a global overview of perspectives for the 21st century. It is a prospect that no nation on Earth can afford to overlook.

Finally, there is clearly a role for universities in the development of an advanced futures discourse and the implementation of foresight. It is a matter of profound regret that so few of them have so far understood how thoroughly the prospects for humankind have altered during the present century, and therefore remain preoccupied with history, boundaries, subject areas and 'knowledge for its own sake'. Yet departments of futures studies or futures research in a number of countries are clearly viable; moreover, many futures courses are taught from within parallel disciplines such as science studies, development studies, geography, sociology and politics.

In summary, the implementation of futures thinking lags well behind its conceptual and methodological development. However, as the practicality and applicability of futures studies become more widely known, this is likely to change.

Critical Futures Studies and the Social Capacity for Foresight

At the human level, foresight is a largely undeveloped human capacity that is, none the less, used ubiquitously in everyday life. At the organisational level, foresight is much rarer, being subsumed into the limited operations of marketing, strategic planning and organisational development. At the social level, a capacity for foresight barely exists at the present time. The great social institutions—government, business, education, commerce—mostly continue as though the particular trajectory of Western culture could continue forever. Specifically, they do not incorporate a clear understanding of the change of scale in human activity and impact that has occurred in recent history, nor of the dysfunctions that have emerged from processes of colonisation and economic hegemony. Thus, as Milbrath and others have pointed out, the old trajectory cannot be maintained. Business-as-usual thinking remains the norm. But we no longer live in 'normal' times. It would be easy to conclude that the outlook is therefore hopeless. But that is not the conclusion of this chapter, nor the collective one of critical futures studies.

At first sight the possibility of seeing the future as anything other than a blank, vaguely menacing space seems unavoidable. Few people have been exposed to futures tools or ideas in their formal education, and few encounter them in later years except in the mostly degraded forms available through popular entertainment and commerce. In the standard view, a common response is to work hard for 'my family', 'my job' or perhaps 'my country', and to stay clear of 'the big picture' because it is too challenging and difficult. In this way, whole populations are deskilled and disempowered. Most young people grow up fearing the future and therefore early on learn the comforts of denial, evasion, avoidance. However, different outlooks are not only possible—some are clearly more desirable.

Critical futures studies began, it is true, in universities.[16] But it is no longer esoteric. Some of the themes certainly require thought (and action), but they are no more difficult to understand and apply than the working agendas of many other fields. Here, for example, are some of the themes that characterise critical futures work:

- the critique of worldview assumptions and practices;
- the reconceptualisation of so-called world problems;
- the review of person–person and person–nature relationships;

- transforming fears and concerns into positive motivation about the future; and
- considering other organising principles for social life, e.g., the notion of a wise culture.[17]

It is clear from the above that critical futures studies seeks to provide a critical purchase on our historical predicament. It does so by attempting to develop, refine and apply tools of understanding and action that, on the one hand, reveal processes of cultural formation and cultural editing and, on the other, reveal options for intervention and choice. In essence, it is this opening-out of personal and social possibility that is so empowering, enlivening and useful.

The critique of worldview assumptions and practices, for example, can be seen as negative and unhelpful to existing institutions and centres of power. But in fact the opposite is the case. An unquestioned commitment to, say, limitless economic growth, or the radically diminished 'ethic' of mass marketing cannot but lead on to an ever less liveable world. It is helpful to know this while there is time to profit from revising the assumptions and practices involved. Similarly, so-called world problems are often no more than the thinly disguised fears of global elites, which bear very little relation to the needs and priorities of the young, the disenfranchised or the poor. Redefining 'world problems' and identifying the sociocultural interests and ideologies that frame them can usefully reveal quite new ways of understanding and responding. The Grameen bank in India is a relevant practical example. Reviewing person–person and person–nature relationships is a useful way of exposing dominant social codes. Are people merely 'human resources', merely 'consumers'? Perhaps there is another diminished ethic at work here. Is nature just an inert 'platform' upon which civilisation is erected? Or is it a community to which we belong? There is a world of difference, now and in the future, between such views.

Fears and concerns about the future are natural and healthy when they alert us to real dysfunctions and injustices. But there are ways of acknowledging such fears and then shifting the focus to strategies of response. I call this the 'empowerment principle'.[18] Finally, the notion of a wise culture allows us to engage in a fascinating process of cultural critique and design. Just how would things change if wisdom was our touchstone, rather than economic convenience or raw technical power? These are significant questions with many practical ramifications for understanding and action in the here-and-now.

I will therefore close this account with a brief summary of a future that could plausibly emerge from the above; that is, from mastery of the four 'layers of capability' underlying futures work in general, and from the wider use of critical futures studies as a distinct methodology. It is obvious that much more work is needed. Positive visions are in short supply, but they are badly needed to provide the young, in particular, with antidotes to the prevailing mood of pessimism and despair.[19]

A Preferred Future

In this preferred future, schools and other progressive social organisations take heed of the innate capacities and needs of human beings according to a broader map of knowledge. The latter extends vertically to embrace a number of ways of knowing so that empirical, communicative and transcendent phenomena all have their place.[20] It also stretches out horizontally and embraces aspects of past and future, thereby greatly enriching the present and clarifying the dense interrelationships between them. Elise Boulding's notion of a '200-year present' is very helpful here.[21]

Then, given a clearly deteriorating world outlook, futures concepts are taken up universally, integrated into many different fields and also developed within an advanced futures discourse. The latter influences other discourses—particularly those of politics, business and education. The change is catalytic. Insights that had been mulled over quietly by perceptive people in many different cultures steadily emerge into the light of day where wider populations can respond to them. The old notion of the future as an empty space fades away and is replaced with a new set of reflexive understandings about the constitution of human cultures and responses in space and time. The future is no longer an abstraction. Rather, a 'grammar' derived from a much wider range of ideas and images becomes widely shared. This strengthens the newly emerging futures discourse. Suddenly the human race begins to grasp the predicament it is in—and the many ways of dealing with it.

Futures tools and methodologies spring up everywhere. A whole growth industry develops as a new generation of consultants, motivational speakers and men and women in all professions and fields begin to adopt, shape and apply these resources in their own lives and work. It is a part of the dynamic 'service sector' that is based on qualitative growth, facilitative processes, communication—and hence involves minimal environmental impacts. The growth of social innovations accelerates and foresight, futures thinking, is implemented just about everywhere. Governments are startled out of their complacency and short-term habits. They are not reformed overnight. But they do ensure that the very best futures thinking is available to them at source. So a new generation of research institutions and IOFs springs up, many sponsored by anxious governments themselves. The international futures discourse becomes truly multicultural and facilitative.

As these social, cultural, organisational and other processes flow together something quite new emerges. It is not the 'noosphere' dreamed of by Teilhard de Chardin, nor the full-blown 'wise culture' sought by visionaries and others. It does not solve all the world's problems overnight, but it does establish a different outlook and perhaps the preconditions of humanly compelling futures. The new quality is a collective capacity for, and commitment to, long-term thinking. A foresight culture therefore emerges at the dawn of the 21st century. It is a culture that routinely thinks long term, takes future generations seriously, learns its way towards sustainability and brings the whole Earth back from the brink of catastrophe.

The old material growth economy is steadily replaced by a 'restorative economy'. Growth itself becomes a dubious concept—unless it is preceded by the term 'qualitative'. Corporations become intelligent, value-based and systems-aware. The earlier commercial outlook disappears and re-emerges in notions of service and long-term quality. Education is transformed. The schools are vital nodes within the new culture, the springboards for society-wide foresight. Universities finally get the message and begin to break down the old interdepartmental barriers: interdisciplinarity thrives. Futures study and research are seen to be one of the emerging disciplines of the new century. A whole new generation of scholars discovers a realm of enquiry that their ancestors would have thought impossible.[22]

The world is no utopia. Wars still break out. Viruses ravage certain areas. It is a nervous time and many species cannot be saved. There is a collective sense of loss and grief. But a different sensibility is abroad. It is one that sees each generation as a link in a chain, not only as inheritors of the past but also as guardians of the future. The species looks out on a newly enchanted world and universe. It grows beyond the primitive ego-states and destructive technologies that drove so much of earlier history. Finally it grows towards maturity.

Notes

1. Gerard Edelman, *Bright Air, Brilliant Fire*, New York: Basic Books, 1992.
2. Richard Slaughter, *The Foresight Principle: Cultural Recovery in the 21st Century*, London and Westport, CT: Adamantine and Praeger, 1995.
3. Hazel Henderson, *Paradigms in Progress*, London: Adamantine, 1994.
4. Richard Slaughter, 'Long-Term Thinking and the Politics of Reconceptualisation', *Futures*, 28(1): 75–86, 1996.
5. See David Hicks, *Teaching About the Future: A Practical Classroom Guide*, Godalming, Surrey: World Wildlife Fund, 1994. Also see Richard Slaughter, *Futures Tools and Techniques, and Futures Concepts and Powerful Ideas*, Melbourne: Futures Study Centre, 1995.
6. Richard Slaughter, 'Critical Futures Study and Research at the University of Melbourne', *Futures Research Quarterly*, 8(4): 61–82, 1992.
7. Peter Schwartz, *The Art of the Long View*, New York: Doubleday, 1991; Jay Ogilvy, 'Futures Studies and the Human Sciences: The Case for Normative Scenarios', in Richard Slaughter (ed.), *New Thinking for a New Millennium*, London: Routledge, 1996.
8. See note 5 above.
9. Richard Slaughter, 'Probing Beneath the Surface', *Futures*, 21(5): 447–65, 1989.
10. Slaughter, 'Critical Futures Study and Research at the University of Melbourne'.
11. See Gerry Mander's reference to the Great Law of the Iroquois in *In the Absence of the Sacred*, San Francisco, CA: Sierra Club, 1991, p. 237.
12. Richard Slaughter, 'From Fatalism to Foresight: A Framework for Considering Young

People's Needs and Responsibilities over the Next 20 Years', Australian Council for Educational Administration Monograph 16, Melbourne, 1994.

13. See Slaughter, *The Foresight Principle.*
14. Gary Hamil and C. K. Prahalad, *Competing for the Future*, Boston, MA: Harvard Business School, 1994; Paul Hawken, *The Ecology of Commerce*, New York: Harper, 1993.
15. Martha Garrett (ed.), *Studies for the 21st Century*, Paris: Unesco, 1991.
16. Richard Slaughter, 'Critical Futurism and Curriculum Renewal', Ph.D. thesis, University of Lancaster, UK, 1982.
17. Some of the themes of critical futures study and of critical futures research are set out in 'Critical Futures Study as an Educational Strategy', in Slaughter, *New Thinking for a New Millennium*, pp. 137–54.
18. The empowerment principle is explored in Slaughter, *The Foresight Principle*, ch. 8: 'Creating Positive Views of Futures with Young People'.
19. See David Hicks and Catherine Holden, *Visions of the Future*, London: Trentham Books, 1995.
20. Ken Wilber, *Eye to Eye: The Quest for the New Paradigm*, New York: Anchor/Doubleday, 1990. Also see Ken Wilber, *A Brief History of Everything*, Colorado: Shambhala, 1996.
21. Elise Boulding, 'The Dynamics of Imaging Futures', *World Future Society Bulletin*, 12(5): 1–8, 1978.
22. See Richard Slaughter, 'Futures Studies as an Intellectual and Applied Discipline', *American Behavioural Science*, special issue on futures studies, forthcoming, 1998.

8
Futures Studies and Futures Facilitators

S. P. UDAYAKUMAR

The basic operating principle of the West with regard to the future is getting there from here without changing anything of what we have here. This status-quoist or establishmentarian view is gripped by the fear of the unknown, the fear of losing prominence and power, and the fear of the 'developing' Other overcoming or overthrowing the 'developed' self. Such a fear psychosis inevitably gives rise to a similar value system: a predictive and positivistic outlook on the future, maintaining order with military might, carrot-and-stick strategies that reward friends and reprehend foes. This Western anxiety and the resultant urge to keep things under strict control permeates every walk of life, including the knowledge industry. Inevitably, history becomes war-oriented, geography strategy-oriented, politics control-oriented, economics profiteering-oriented, the media propaganda-oriented, education drilling-oriented and so on and so forth.

It is no wonder then if the singular 'future studies' is all about sidelining the multiple streams of non-Western imaginations, fantasies, preoccupations, energies and endeavours, and streamlining the Western project as the unproblematic and all-encompassing singular channel that flows towards the whole of humanity's 'life, liberty and the pursuit of happiness'. Imbibing the common tenets of the related 'disciplines' of the Western knowledge system such as 'engineering' or 'management', 'future studies' shuns anything and everything that does not subscribe to Western scientific and objective rationality. In this disciplinary quest for academic initiation and popular recognition, the plural futures are reduced to a synthetic singularity and institutionalised with little spontaneity. The first casualty in this development is, of course, human ingenuity.

The preponderance and power of institutions and institutionalisation is so intense in the Western scheme of things that the chasm between micromotives and macrobehaviour is quite tremendous, and deliberately so. For instance, 'What goes on in the "hearts and minds" of small savers has little to do with whether or not they cause a depression.' Conversely, economists deal with 'systems that lead to aggregate results that the individual neither intends nor needs to be aware of, results that sometimes have no recognizable counterpart at the level of the individual'. This kind of segregation is quite complex in relation to the dynamics of individual choice. However, 'One might even be tempted to suppose that some "unseen hand" separates people in a manner that, though foreseen and intended by no one, corresponds to some consensus or collective preference or popular will.'[1]

Many futurists have clearly elucidated that such an 'unseen hand' does exist in the contemporary Western academic discipline of 'future studies'.[2] The inherent micro–macro segregation helps Western futurism marginalise the local and human-oriented non-Western streams of future as much as the internal streams of dissension as irrational and wacky, ethnic, magic etc. Furthermore, it holds the broader systemic predicates, such as science, state, globe, market, as the most legitimate and meaningful sources of futurism. Despite all the grandiose claims of individualism, democracy, freedom, human rights and so forth, humans' control is compromised for the control of humans. Micromotives are doubted and macrobehaviour is demanded.

One has to choose between a synthetic macro-future and organic micro-futures. Any attempt at liberating 'future studies' from the Western academic disciplinary trappings and discourses should begin by encompassing all possible ways of meaning production in life, and of measuring success and happiness on the Earth. The post-colonial futurists cannot continue to be just the agents of conscientisation marked by political dissension; they also need to assume the additional role of rescuing human communities, both large and small, and their rightful futures. The emerging 'futures studies' becomes a rescue mission. Developing the concept of 'futures facilitators' who efficiently lead and guide individuals' and human communities' forward journeys, this chapter attempts to show how the larger schemes tend to swallow up local dreams. Focusing on scientism and globalisation that seek to promote macrobehaviour, the chapter dwells on the need to emphasise micromotives in facilitating human futures and in designing the devices to study them.

The Futures Triumvirate

The future being multidimensional, any attempt to approach it from a religious, scientific, economic or other such limited perspective can only result in partial understanding. Even a sophisticated combination of all these and other dynamics will still be deficient, as the human element that forms the core of the future defies sweeping generalisations. However, science and religion have often claimed to have a better grasp of human futures.

Science, according to Dennis Gabor, the British physicist who won the Nobel prize in 1971, 'means the application of man's reasoning power to ways for achieving his aims'.[3] Gabor divides humans into 'cyclic' and 'acyclic' men, or 'common' men and 'dedicated world-improvers'.[4] According to him, 'the common man may be less used to balancing complicated rational arguments against one another than the highly educated man'.[5] On the contrary, 'Science sprung first from the urge of exceptionally gifted men to exercise their mental powers.'[6] Gabor contends that 'a stable but progressive society can exist only if the "cyclic" type is common, and the other is rare'.[7] In this society, '*True* science will provide congenial work for uncommon men and women as far as thought can reach.'[8] However, there is one predicament for 'the social inventor' in this process, and that is the need for 'the engineering of human consent'.[9]

Extrapolating this view to the larger world, Gabor posits:

> Though modern research has shown that Indian, Chinese, and Japanese science often went parallel to European thought and sometimes preceded its discoveries, it is still justifiable to consider the Greeks as the fount of our science, because Eastern science remained unknown in Europe at the time of the great break-through in the seventeenth century, in the epoch of Descartes, Galileo, Kepler, and Newton.[10]

Seeing no need even for 'engineering' any consent at the global level, Gabor turns to the future and argues that 'The future cannot be predicted, but futures can be invented.'[11] He aspires towards 'a future world in which people will sing and laugh spontaneously when they go to work, and when they return from work. A world in which there is much happiness for the common man, and much creative struggle for the uncommon man.'[12] Gabor proposes that a group of 'very smart people' be required to prepare a list of possible futures, and that society be asked to decide democratically which of these futures corresponds closely to the sort of world we would want to live in. And then we could design policies intended to get us there.[13]

Following the Gaborian reasoning, Michael Shermer arrives at similar answers to his question of 'why people believe weird things'. He contends that 'Scientists and secular humanists' have good answers to the existential questions such as the meaning of life and ethics; however, these answers have not reached 'the population at large'.[14] According to Shermer, 'Scientific explanations are often complicated and require training and effort to work through. Superstition and belief in fate and the supernatural provide a simpler path through life's complex maze.'[15] So people have trouble distinguishing science from pseudoscience, history from pseudohistory, and sense from nonsense.[16] Equating the 'uncommon man and Common man' with 'scientists and futurists', Robert Park, a physicist, argues that scientists plot a new course for humanity with each hard-won insight, but futurists go for astrology and 'tell people what they want to hear'. If we follow the contours of the scientific terrain and do not focus on too distant a goal, Park concludes, 'the futurists are simply irrelevant'.[17]

The worldview of these scientists is obvious. There are two forces in the world. On the one side, there are these 'very smart' 'international brahmins' who have the innate knowledge and wisdom and the naturally ordained duty of educating and guiding us all to salvation. On the other side, we have the 'common' folks with a bunch of futurists and other such 'crazies' who cannot help themselves but will resist these scientocrats' scheme of taking us all to a 'Brave New World.' Although they do not quite catch what futures studies is all about or who futurists really are, they do understand that resistance to domination is a basic tenet of both futures studies and the larger human society. Furthermore, it is indeed a compliment that futurists who are devoid of the chosenness syndrome or 'manifest destiny' thinking are identified with the 'common men and women' struggling together for common futures. The political nature of futures studies is acknowledged and feared.

Although futures studies has its own rationale and methodologies, it does not appropriate the burden of envisioning human futures for itself. The democratic pluralism that futures studies insists on indeed helps dissenting voices flourish and proliferate. If the future is everybody's fate, predicting or envisioning has to be everyone's forte. And science has to co-exist with scores of other ways to tease out the probable and to dream about the possible. As Carol Tavris argues, pseudo-scientific predictions are more appealing for some people than sober scientific predictions because the former speak in reassuring tones of certainty, and they are based on intuition, faith or individual experience. On the other hand, scientists speak in the language of probability; they predict with great accuracy that smoking will increase the likelihood of early death, but they cannot predict if any individual smoker will have lung cancer. Their reliance on objective evidence may make their predictions accurate but they seem tentative, remote and encumbered with qualifications.[18]

Answering the question of why people cling to ideas society judges to be outdated, Anthony Aveni posits that 'Either we long for an imagined, more beneficent world that we feel has been lost to us or we shun the invasive and personally threatening philosophies that gave rise to our scientific way of explaining nature, or both.'[19] Our *zeitgeist*, Aveni contends, is the desire to unravel the design and purpose of the universe with the help of science. As the particle accelerator and space telescope are to penetrate the divine plan of creation, the computer becomes the scientists' 'philosopher's stone' and their goal is 'the simulation of all possible future versions of ourselves'.[20] At the other end of the spectrum are magical acts that cover a whole array of varied practices around the world.

> Each magical act is really a wish—an attempt to persuade a charismatic cosmos [a human being] is confident [s]he can talk to. Magic is an attitude that makes you aware that you must act accordingly. It is a form of activism and participation as sensible and practical in many cultures of the world as scientific trial-and-error testing is in our own.[21]

Claiming that both oracle and computer put a share of the power of God in us, Aveni posits:

> If divining means giving a sign of future things to come, then magic's role in society is to imbue the future's possible course with that wished-for element.... Believing in a participatory cosmology spurs us on to action. It leads to the notion that our actions have a real stake in what will happen in the future. We really *can* affect what happens in the world.[22]

As science and religion lock their horns so inextricably, the force that walks away with the victor's trophy, however, is (finance) economy. The elimination of barriers to trade and capital and the rise of communications technology have created a global financial market-place that is dominated by huge financial institutions 'with short investment horizons'.[23] This 'naive globalism', as Robert Kuttner calls it, includes a few basic precepts: that free competition is good nationally and even better globally; that liberal capitalism would be self-regulating with a few basic ground rules; that the true form of capitalism entails a minimal role for the state; and that markets should be transparent and porous, giving a free hand to the investors. This globalism intensifies the aspect of capitalism—financial markets—that is most vulnerable to herd instincts and damaging irrationalities.[24]

The ongoing secretive efforts of the global corporations and wealthy investors to draft what the director of the World Trade Organization calls 'the constitution of a single global economy', the Multilateral Agreement on Investment (MAI), would be the penultimate achievement of globalisation. Giving investors a legal status on a par with that of nation-states, the planetary regime will require that almost all sectors of national economy be open to foreign investment. Having freed themselves from government regulations and social obligations, the corporations will force citizens and their governments to take huge risks but will retain the right to sue governments and claim monetary damages.[25]

As Lester Thurow puts it, 'Like Columbus and his men, all of us aboard the good ship "capitalism" are sailing into a new uncertain world.'[26] Columbus seems to be hailed rather deliberately here because there is an unmistakable Columbusian element in this whole scheme. The global market, hailed as 'the dominant, worldwide force of the early 21st century'[27] assimilates everything[28] into its ambit. Admitting that '[i]n capitalism there is no analysis of the distant future',[29] Lester Thurow presages that 'capitalism is going to be asked to do what it does least well—invest in the distant future and make deliberate adjustments in its institutional structure to encourage individuals, firms, and governments to make long-term decisions.[30] If the flaws in the global capitalist system are not recognised in time, according to George Soros, 'the deficiencies are likely to make their effect felt, and the boom is likely to turn into a bust'.[31]

The triumvirate that we discuss here—science, religion and (finance) economy—have broader implications. Formulating the axiomatic Western 'three-in-one' product of nation-statism, scientism and developmentalism, science actually symbolises the organisational resources at the national level. Religion/magic represents the inner resources of individuals and local cultures that lie buried underneath the national scheme. The glossy capitalistic 'all-in-one' remedy epitomises the material resources at the global level. Which of these forces does facilitate human futures more efficiently—inner resources that spring from the core of an individual and organic local human collectives, or the organisational resources that are imposed by the dominant mode of the national society, or the material resources that are commanded and controlled by a few global actors? Any answer other than the material resources at the global level is considered today naive and even reactionary. Although material resources are an important part of life on the Earth and human futures, they are not the only thing that we ever need here. Before answering the above question in earnest, we need to understand the concept of futures facilitators.

Futures Facilitators

Writing 'On Memory', Jerome K. Jerome (1859–1927), a popular English humorist, likens it to ghosts that are always with us day and night. Much like a haunted house, memory's walls are ever echoing to unseen feet. Pleading to do away with vain regrets and longings for bygone days, Jerome contends that 'Our work lies in front, not behind us; and "Forward!" is our motto. Let us not sit with folded hands, gazing upon the past as if it were the building; it is but the foundation.'[32] In order to illustrate his point that we need to press ahead despite 'the shadows of our own dead selves', Jerome relates the childhood story of a brave and resolute knight:

> Now, as this knight, one day, was pricking wearily along a toilsome road, his heart misgave him, and was sore within him, because of the trouble of the way. Rocks, dark and of a monstrous size, hung high above his head, and like enough it seemed unto the knight that they should fall, and he lie low beneath them. Chasms there were on either side, and darksome caves, wherein fierce robbers lived, and dragons, very terrible, whose jaws dripped blood. And upon the road there hung a darkness as of night. So it came over that good knight that he would no more press forward, but seek another road, less grievously beset with difficulty unto his gentle steed. But, when in haste he turned and looked behind, much marvelled our brave knight, for, lo! of all the way that he had ridden, there was naught for eye to see; but, at his horse's heels, there yawned a mighty gulf, whereof no man might ever spy the bottom, so deep was that same gulf. Then, when Sir Ghelent saw that of going back there was none, he prayed to good Saint Cuthbert, and, setting spurs into his steed, rode forward bravely and most joyously. And naught harmed him.[33]

Although we need to face the road ahead with forward-looking strategies, that alone does not ensure a smooth ride to the futures, as the road is often paved with inherent uncertainties and enormous anxieties. When the journey has more encumbrances, we are in need of futures facilitators just as Sir Ghelent turns to Saint Cuthbert.

Futures facilitators are neither completely tangible, like physical concepts or chemical catalysts, nor totally elusive, like magical powers or spiritual forces. Being a vague amalgam of these two extremities and a whole array of intermediaries in between, futures facilitators take shape depending largely upon the inclinations, idiosyncrasies and interpretative and creative faculties of the entity concerned. As people or communities struggle to overcome difficulties and obstacles and to push their life forward, they look for two things: 'futorcism' that will get rid of 'the shadows of our own dead selves', the ones that we ourselves create and the ones that others impose on us; and 'futurism' that will facilitate a smooth transition to the desired futures. They turn to any number of religious and secular sources of strength—what we call 'futures facilitators'—that combine both of the abovementioned services quite effectively. As William Young puts it, 'Historically you can see that the most intense interest in and the most vivid imagery of heaven tends to come at times of persecution. Focusing on heaven and what God has in store becomes a very hopeful way for people to endure suffering.'[34] Even if no particular hardships are involved in the current journey, faith in futures facilitators promises to regulate the chaotic mystery that lies ahead and to bring forth only the best. Futures facilitators work exactly like astrologers who help clients through anxious moments 'by offering them confidence and bolstering their will to make decisions about the kinds of everyday affairs that still trouble us'.[35]

As a result of this inner dynamism and the supposed terrestrial brevity, there come more and new initiatives and actions endowed with hope, possibilism, creative energies and talents. As a result, there is a stronger sense of futurefulness. Besides fulfilling the basic requirement of fostering life and perpetuating it just as future itself does, futures facilitators need not fit into a fixed criterion any more than human volition. For instance, how would one counter the claim that the 'promised land' is in cyberspace[36] or other such assertions of the 'Black Futurists in the Information Age'.[37] Thus, there are no 'good' or 'bad' futures facilitators but only those whose nature is either enhanced or depraved, leading to varied ends with varying results.

Although futures facilitators may not be neatly classified or quantified, we can see them at work at all different levels in human society: individual, family, immediate community, larger communities and globally. The futures facilitators are most effective and powerful at the individual (micro) level, as they are deliberately chosen by the inner voice/vision and cherished with faith and conviction. At the intermediate (meso) levels such as clan or country, the futures facilitators are more of a group trend rather than a personal insight. Examples could include the influence of a social reformer, a shared economic interest and a political revolution. Apart from the notion of humanity and some shared humanistic values, there have been a few aberrant futures facilitators at the global (macro) level such as the well-meaning linguistic attempt of the Esperantists or the ill-meaning political schemes of various actors and ideologies. The schemes of crusaders, colonialists, communists, Nazis, fascists, imperialists, Cold Warriors, ethnic cleansers and capitalists cannot be considered as futures facilitators, since all of them have taken a huge toll on human lives.

Futures facilitators flow in all different directions; they are interpreted and understood differently in different contexts; and they may be shared by both individuals and larger human communities. For instance, Islam inspires scores of societies around the world as a way of life and it also guides pious individual Muslims. Similarly, Ram, the favourite personal deity of millions of Hindu Indians, inspires family futures, larger collective identities, and even some destructive political manifestations. At the broader levels, futures facilitators are of a more imposed and manipulated variety, which may be vague or concrete, like 'our way of life' or 'national founding fathers'. In the final analysis, however, futures facilitators do look vivid for many and help them to see ahead and go forward. It is indeed worthwhile to consider the functioning of the futures facilitators in all the three streams of futurism that we discussed earlier.

Spiritual Local Futures

In the case of individuals, the futures facilitators get even vaguer and the vagaries even more quixotic. John P. Parker (1827–1900), a prominent and daring African-American conductor on the Underground Railroad, is a telling example. In both the early part of his life, when he was desperately trying to win his freedom, and the later part of his life, when he helped hundreds of his people to achieve freedom, he relied on several factors such as his rage ('I was like a mad bull hitting out in every direction at my enemies'), 'confidence in myself, in my ability to meet any and all situations which might arise to confront me', 'crazy luck' and 'good fortune'. He gave credit to his 'active mind', which constantly worked out imaginary plans of escape and formulated a theory that a plan and the timing of its execution were quite crucial. The same brave and planned Parker also attributed his adventure to external sources such as receiving 'a message from the high heavens', and his way opening up at crucial moments.[38]

Even the widely prevalent futures facilitators such as religions stand altered by the personal stamp of one's own life experiences and expectations. Former US President Jimmy Carter asserts that he prayed more during the four years of his Presidency than at any other time in his life 'for patience, courage and the wisdom to make good decisions. I also prayed for peace—for ourselves and others'.[39] He contends further that his 'faith as a Christian' has provided the necessary stability in his life and that 'to have faith in something is an inducement not to dormancy but to action'.[40] This kind of religious/spiritual faith leads to many supernatural beliefs such as boons, blessings, visions, darshans (holy audiences), simulacra (perceived simulations of images), feng-shui (Chinese geomancy) and other such metaphysical assertions. Almost every single culture on the Earth has its share of such beliefs and practices.

There is also an ample number of non-religious incentives to perpetuate the futures; however, they are often connected to the spiritual plane. The 'basic goodness' that Oseola McCarty, an 88-year-old poor laundress in Mississippi, exhibited by donating her hard-earned $150,000 savings to the University of Southern Mississippi for scholarships opened the world for others and in the process got her own world opened up also.[41] The unclothed and badly burned 9-year-old Thi Kim Phuc crying and fleeing from the scene of napalm attack in Vietnam ordered by an American commander in June 1972 laid a wreath at the Vietnam Veterans' Memorial in Washington, DC in an act of reconciliation and forgiveness. Phuc, who lost two younger brothers and had a third one severely burned in the attack, told her American audience: 'Even if I could talk face to face with the pilot who dropped the bombs, I would tell him, "We cannot change history, but we should try to do good things for the present and for the future to promote peace."'[42]

If giving and forgiving facilitate futures, so do remembering and reviving hopes. Maria Rosa Henson, who was imprisoned by the Japanese Army in a brothel in the

Philippines during the Second World War as a 'comfort woman', was forced to have sex with between ten and thirty Japanese soldiers a day. Overcome with shame and guilt, she was severely traumatised for months afterward. During this period, she tried hard not to lose her sanity: 'I tried to work my mind all the time, night and day. I did not let my brain stop thinking. Always thinking and thinking. I learned to remember everything, to remember always, so that I will not go mad.'[43]

Just as Henson remembered things to preserve her sanity, Jacob, the protagonist of Jurek Becker's novel *Jacob the Liar*, fabricates things to sustain his group's sanity. He is a forced labourer in a Nazi ghetto that is completely cut off from the rest of the world. One night Jacob is summoned to the German military office for having broken the curfew. Before he is miraculously released, he overhears the news broadcast of the Red Army's advance to a town some 300 miles away. Jacob tells this great news in his ghetto but it is not believed; so in exasperation he claims to have heard it on his hidden radio. From that point onwards people constantly pester him for further news and Jacob begins his daily fabrications about the relentless Russian advance. Providing the much-needed hope that has been absent in the ghetto, Jacob's refreshing reports from his non-existent radio bring down the suicide rate to zero.[44]

Quite often we find ourselves muttering the words of Jerome K. Jerome: 'I am alone, and the road is very dark. I stumble on, I know not how nor care, for the way seems leading nowhere, and there is no light to guide.'[45] In those moments of darkness and futurelessness, we turn to many different sources for light and guidance. Any such source that illuminates the future and elevates life is indeed a futures facilitator. The futures facilitators help one endowed with the futures drive overcome the present obstacles and accomplish the desired futures by providing the much-needed psychological impetus, spiritual hopefulness and overall atmosphere of magical optimism. However, the process remains very individualistic and renders sweeping generalisation impossible.

These mostly external sources of inspiration that secretly bring about the internal transformation, determination, favourable environment and inevitable positive result get the credit for the accomplishment. Quite often, the religious/spiritual tone is rather strong in individuals' cases. Even when the eventual outcome is partial or mixed, it is often positively interpreted as a 'blessing in disguise' or 'God's will'; or, as a Tamil proverb puts it, 'the arrow that came for the head went off with the turban', and so forth. Thus the futures facilitators never fail. When a quirk of luck caps the event, the power of the futures facilitators blooms in full exuberance. So psychological factors, discursive interpretations of the situations and a quirk of timeliness/luck/fate complete the powers of futures facilitators.

Preparing for the future is a deliberate act of self-determination, which one undertakes with self-assurance derived from sources chosen personally at one's own free will. The self-affirmation of one's own futures and letting others do the same for themselves are complementary and even interdependent processes. Dream and let dream! If the self-affirmation of one's futures gets out of hand to the point of considering oneself or one's futures vision as chosen, the complementarity and interdependence of human futures gets misunderstood and we encounter 'civilising missions', 'discovery' attempts and 'manifest destiny' projects. When someone endowed with an innate arrogant assumption that others are inherently flawed or inadequate and therefore could not possibly engage their own future unilaterally assumes that responsibility with hidden selfish motives, such an act could be considered futures facilitation or futures colonisation, depending on one's politics.

The bearing of futures facilitators on politics, however, is the assertion of faith in human agency, and affirmation of hope and futures. Underscoring everyone's unique importance in life, futures facilitators become a statement of democratic pluralism, a celebration of both the oneness and the diversity of humanity, an assurance of people's right to choose and dissent, and, quite simply, the state of being free. Although unfettered facilitation of individuals' futures is an important precondition for the collective futures to be free, quite often the latter tends to impose an artificial uniformity on the former in order to streamline the collectivity. Even a cursory look at the contemporary national scheme and the impending global scheme will illustrate that claim.

Organised National Future

The foundational futures facilitator in the present-day world that blends human organisation, orientation and destination is the 'Holy Trinity' of nation-statism, scientism and developmentalism. When the European settlers or colonisers violated indigenous peoples, the first thing they invariably did was to sap their victims' strength: usurping the land, pillaging the resources and breaking in on the traditional customs and beliefs. Having succeeded in making the natives feel weak and worthless, the intruders imposed their own values and ways of life on their victims. The indigenous models and methods, which did have their own share of faults and problems, were interrupted and an alien system imposed in their place. Consequently, many peoples around the world are stuck in Eurocentric schemes of community organisation, identity construction and meaning production. The production of locality for humans has come to mean the production of subject-citizens of nation-states. Vague as they may be, these 'imagined communities' (Benedict Anderson) have come to look vivid for many and to help them visualise their singular national future.

The spate of partial political decolonisation of the last few decades has changed the scene a little, but the overall game continues to be played the colonisers' way, with their rules and strategies. The power elites in the 'Third World' accept the colonisers' models and try to imitate them by turning a blind eye to the indigenous value systems. Most of the newly independent countries have made a conscious decision to pursue the Western model of nation-building, which requires a particular economic model, a standing army, strategic industries and an extensive bureaucracy to direct and shape 'national development'.

Providing the overall orientation to this infamous development project, science and technology mould the nation-state and its existential foundation, the military establishment. A 1945 *National Geographic* article, entitled 'Your New World of Tomorrow', boasted 'how science is remaking our world' through inventions, discoveries, devices and products that 'were developed for our Army and Navy under the urgent demands of war'.[46] Under the seemingly innocuous and even philanthropic spirit of scientific and technological enterprise and allurement, there lies an ideology and programme. As an American academic attests in his survey of the Second World War period:

> The most significant discovery or development for science and technology to come from the war effort was not the technical secrets that were involved in radar or the atomic bomb. It was the administrative system and set of operating policies that produced such technological feats.[47]

Adding to this contention, a top American politician claims that the 'organizational innovations developed on a national scale' have made the benefits of science possible, provided military strength, and, above all, given form and character to the American society.

Lamenting Western Europe's technological shortcomings compared to the United States and Japan, a European author points out that the problem is not technological in character but 'political, economic or social impediments to innovation' and that the remedy for that is in public policy.[48] So the purpose of policy should be to influence the likelihood of chance occurrences between fundamental discoveries and practical interests

> by increasing the density of both kinds of activities and the velocity of the circulation of ideas and problems from both areas of activity in spaces which ensure interaction. Increasing the density is a matter of investment, velocity is the result of entrepreneurship, and creating properly enclosed spaces is the task for organization.[49]

In other words, policies to support the development of science and technology will contribute substantially to industrial innovation with the creation of an overall economic environment. An appropriate balance between 'technology-push' and 'market-pull' policies is called for.[50]

Inevitably, the 'national development' spearheaded by science and technology is seen as a higher standard of material life and transformed into a series of aggregates such as 'raising the Gross National Product, assuring a certain rate of growth, and in turn fulfilling a series of production functions, consumption functions, utility functions and other "principal components".'[51] Viewing development in terms of GNP, the so-called modernisation approach emphasises providing markets for manufactures in international competition, warranting large investments in industrial growth and infrastructure, providing the basis for diversification and creating a critical mass in technical personnel and investment resources. For a predominantly rural 'Third World' country this means a necessary shift from a rural structure to an urban one based on large-scale industrialisation, compromising on the agricultural sector and small industries. As Pradeep Bhargava points out, those sectors of the economy where the capitalist mode of production had been adopted emerged as 'a nucleus of economic activity and political power' and the other sectors, which continued with the pre-capitalist mode of production, were pushed to the periphery of the nucleus.[52]

While the middle class and other dominant classes of industrial bourgeoisie and capitalist farmers vie with each other to dominate the regime of nation-states, 'a great majority of the population in the periphery has been excluded from institutional participation in the political process especially in the non-democratic power structure'.[53] The subaltern classes, which include the urban poor, landless rural workers, marginal farmers, deprived ethnic groups, women and children, have no leverage whatsoever in policy decisions or development activities, as these are all top-down projects. Thus the nation-state, assisted by science and development, provides the organisation with spatial orders, linear time and rational regulations that are based on beliefs rather than insights. Since this is the overall national framework all over the world, we easily fall for the conventional political socialisation and invoke our national identity, the scientific achievements of our country, our nation's military might and so forth. Organised science provides the future orientation for the state to reach its desired destination of control.

Monetised Global Future

One of the civilian fall-outs in the above scheme is consumerism and hedonism, the tip of the emerging globalisation iceberg. This innocent-sounding term, 'globalisation', evokes some specific images in one's mind: that it is an altruistic geographical notion that attempts to integrate the 'globe', that the integration process consists of both centrifugal and centripetal movements, that it rotates on its own moral axis and revolves around the notion of universal peace and justice, and that this 'chance discovery' of humanity has had no past and is completely devoid of the political. Most of the discussions on globalisation indeed reflect this naive understanding. They invariably applaud the flow of capital and investment, information technology, communications, economic transactions, political values, cultural norms and the overall orientation in life. The strict North–South one-way traffic, calculated detours from discrete places, greedy rush to the greener pastures and the preservation of the same old colonial traffic rules go grossly overlooked. In fact, as Francisco Sagasti sums up forcefully,

> [W]e are witnessing the emergence of a fractured global order—an order that is global but not integrated; an order that pulls all of us in contact with each other but simultaneously maintains deep fissures between different groups of countries and people within countries;

an order that segregates a large portion of the world's population and prevents it from sharing the benefits provided by scientific advances and technological progress.[54]

Endowed with capital, technology and political power, the capitalistic North promotes globalisation as the salvation theology for the modern world. It fails to discern or discuss globalisation giving rise to a future-blind attitude towards the Earth and its resources, consolidating the wrong and lopsided development of the world, destroying local industries and indigenous expertise, tying the welfare of the whole world to the operation of a single economic scheme that is dominated by a few individuals and corporate groups, creating the danger of overdependence on foreign investments and imports leading to outside control, and pushing humanity to a desperate 'do *and* die' situation.

The power elites of all different nation-states, having been functionally dependent on Northern powers and their capitalism, are linked with the global market. One of the outcomes of this joint venture is growing reliance of the former on borrowings from the latter. Giving rise to debt-servicing, which consumes a substantial portion of export proceeds, the borrowing also obliges the elites of the South to adopt externally imposed structural adjustment programmes, the 'civilised' version of human sacrifice. These conditionalities normally include across-the-board price increases, import liberalisation, raising of utility charges and so forth. Fuelling inflation and increasing hardships, the structural adjustment programmes hit the subaltern classes hardest and most directly. Large-scale export production hits the small producers and farmers; labour rights get overlooked; women and children become more vulnerable to further exploitation; the poor get poorer and the rich get richer.

After all, globalisation means Westernisation means Americanisation means monetisation. While humanity seems to be poised, at the superficial level, for greater solidarity with increasing travel and enhanced communication, what is also happening concurrently is an influx of covert Christian missionaries, industrial spies and underground elements such as arms dealers, drug traffickers and sexual offenders in the guise of English teachers, aid workers, traders, tourists and researchers into the South. Although the so-called globalised world is made to sound efficient and glamorous, it does not even claim to create a future that transcends the faults of the present.

Rescuing Futures and Reinventing Futures Studies

A comparative analysis of the micro, meso and macro futures facilitators points out some interesting dynamics. As the size of the human entity gets larger, the futures facilitators get more and more distant and impersonal, and their facilitating potential stands dissipated. While the desired futures at the micro level are wilful and hand-picked, the chosen futures at the larger levels are often imposed. Moreover, the political deliberation in selecting one's favoured futures is much stronger at the micro levels because the personal stakes are quite high. None the less, the contemporary trend is towards nationalising and globalising human futures. So any attempt at rescuing all our futures must involve resisting nationalised and globalised futures by challenging the imposed spatial arrangements, time-orders and herd instincts and restoring the right and power to decide one's own futures. Such efforts call for slackening of the nationalist and naive-globalist strait-jackets worn around the world and the strengthening of human solidarity both at the tangible local level and at the spiritual global level by struggling for an equivocality of being a local-citizen, national-citizen and world-citizen and adding interesting dimensions to the discursive space enveloped by the identity–meaning–futures mix.

However, Aijaz Ahmad warns us that the coercive nature of the national identity cannot, in fact should not, be opposed by invoking local communities as sites of autonomy and harmony. The undue romanticism of invoking these autonomies is, in most cases, quite misplaced, because these local communities are, as a norm, structured hierarchically, mainly on the axis of caste, and almost always with unequal division of labour and assets. Also these local communities often rest, structurally, on the patriarchal household. Most often the local communities look idyllic or like an alternative to the nation only from the viewpoint of the upper caste, the patriarchal household and the ownership of agrarian property. On the other hand, the autonomies of local communities are being blown apart by the process of state formation and market liberalisation, and these autonomies can be defended 'only by struggling, simultaneously, for a kind of national project that would protect these communities from the destructiveness that comes spontaneously to states and markets when they are left to their own devices.'[55] We can do just that by turning to the local histories and futures; such multiplicity of histories and futures may enhance the political logic of democratisation that requires more and more autonomous decision-making at local and regional levels.[56]

Such an equivocality will render unidirectional linear journeys towards larger schemes and themes (nationalism and naive globalism) deleterious, and require 'oceanic circles' (Mahatma Gandhi) forming and broadening all over. At the centre of the circles will be, of course, individual human beings, who breathe faith and hope with their own respective futures facilitators, transcending borders and transacting meanings. The argument here is not to relegate human agency and volition to some mysterious and amorphous inner constructs that stick to the monotheisms of religion or science or finance economy, but to regain the right and power to configure one's own futures in accordance with one's internal cosmos and external galaxies. Anything that makes the futures open and possible for a human being is indeed a boon for that person and to the whole of humanity. Whereas making the necessary ingredients—spiritual guidance, scientific and technological aids, material resources and so forth—available is the larger humanity's responsibility, mixing them in the right proportions should be individuals' or local communities' concern.

Given the mostly deleterious effects of absolutist identities, one should foster multiple, fractured and mutually inclusive localities such as local-citizen and regional-citizen, among others, and realise that the production of all these localities need not be cogently reported.[57] Similarly, we can no longer neatly partition the world into confining categories of religion and science, tradition and modernity, or globalism and protectionism, and take an unyielding stand for one or the other. For instance, 'we are long past the point of debating whether to live in some kind of pervasively technological society; the real question is what kind of technological society do we wish or can we afford?'[58] Such a realistic idealism does not necessarily end science or trade or futures but it begins to look at them from various angles at varying moments.

As it is clear now that the whole world cannot afford to become cities and to adopt the consumerist lifestyle; we should adopt alternative lifestyles. Instead of surrendering to the scheme of the powerful and blind, and subjecting ourselves to further exploitation, we should resist and struggle to save all of us. Leave alone the impact of globalisation on our living environment, on the chasm between the haves and the have-nots, and on the socioeconomic-political realm; the worst crime committed is the complete negation of the underprivileged and the unwilling. Profiteering globalisation assumes an unproblematic, unanimous representation of all humanity, and feigns a singular vision and voice. This 'rich man's burden' wipes out the sporadic sustainable development efforts and imperils the two-thirds of humanity who live in about two million villages. Individual communities around the world should selectively endorse or boycott products and services, and formulate other strategies to consolidate their interests, depending on local contingencies. Forming transnational solidarities with our counterparts all over the world, we should promote communication among villages, and between villages and cities. Demonstrating our belief in pluralism, and respect for the choices of others, we should continue to raise our concerns on the issues that imperil our interests and the Earth.

Although giving disciplinary trappings to the study of human futures will inevitably play into the hands of Western academia because of its sheer power and prominence, giving the effort some kind of form and content is important. To think along the lines of Kaoru Yamaguchi, any effort at liberating futures studies should accommodate academic, semi-academic and popular discourses, and give as much opportunity to sophisticated industry forecasts as, say, to indigenous peoples' futures predilections.[59] An excellent example is the project that asked the children of Hawai'i to describe the future they would want to live in. Having collected the answers, the 'Hawai'i Benchmarks: A Voyage to the Future' project worked with the community to improve the quality of life in Hawai'i by articulating a shared statewide vision for the future, identifying goals and measurable benchmarks against which progress towards this vision could be plotted, and focusing public and private action on key benchmarks. The project's driving question was this: 'While there are many individual visions for the future, can we find in the desires of our children, common elements of a future that we can all share?'[60]

Despite this kind of openness and creativity, the refurbished futures studies would require the intellectual leadership of the post-colonial futurists, who should lay a categorical emphasis on the aforementioned political positions and normative principles. The reorientation of futures studies should be basically a rescue effort. To rescue is to resist. The resistance should be both basic and far-reaching, challenging all of the Western foundational assumptions, such as Rousseau's famous educational dictum: 'To produce a man or to produce a citizen?' Futures studies should not seek to 'produce' anything out of anyone but only strive to clear the hurdles and create the conditions in which individuals and communities can bloom into full exuberance. Futures studies should create and perpetuate socio-economic-political spaces that allow the free and unfettered flow of individual futures into collective ones at the local, national, regional and global levels and the collective ones to flow back to individual futures in all possible ways and means and directions. All these and other such efforts should turn futures studies into an effective futures facilitator.

When all is said and done, 'two obvious possible alternatives confront the human race today',[61] and both are equally bad. One is atomic annihilation of the world, and the other is 'avoidance of war through a unification of the World under a dictatorial world regime in which the rich and powerful minority would league together, all over the World, to hold down the poor and backward majority.'[62] With this diagnosis, Arnold Toynbee asks the younger generation to find a middle way, that is, 'some form of world unity that will enable the human race to survive and prosper without oppression and without reaction'.[63] Ultra-nationalism with nuclear danger is oppression; globalisation is reaction. To fight oppression and reaction, Toynbee offers two specific pieces of advice. First, the whole thing has to be done in the 'Gandhi spirit' without hatred and violence. Second, we should not be discouraged or embittered by the smallness of the success that we may achieve in trying to make life better. It is important to remember the Buddha's teaching that the load of *karma* resulting from bad action can be diminished and thrown off by good action. The requirements are, of course, 'the gentleness, the patience, the long-suffering'[64]—the common traits of futures facilitators!

Notes

1. Thomas Schelling, 'Micromotives and Macrobehavior', in Robert Ellickson *et al.*, *Perspectives on Property Law*, Boston, MA: Little, Brown and Company, 1995, pp. 478–9.
2. For more discussion, see Ziauddin Sardar, 'Colonising the Future: The "Other" Dimension of Future Studies', *Futures*, 25(2), March 1993; 'Ziauddin Sardar Interviewed', *Metamute* (online magazine), 1, 1998; James Dator, 'Decolonizing the Future', in Andrew

Spekke (ed.), *The Next 25 Years: Crisis and Opportunity*, Washington, DC: World Future Society, 1975.

3. Dennis Gabor, *Inventing the Future*, London: Secker & Warburg, 1963, p. 184.
4. Ibid. 129.
5. Ibid. 189.
6. Ibid. 105.
7. Ibid. 129.
8. Ibid. 158.
9. Ibid. 185.
10. Ibid. 105–6.
11. Ibid. A recent book of predictions by a physicist is worth mentioning here: Michio Kaku, *Visions: How Science Will Revolutionize the 21st Century*, New York: Anchor/ Doubleday, 1997.
12. Gabor, *Inventing the Future*, p. 128.
13. Robert L. Park, 'Future Schlock', *New York Times*, 30 December 1997.
14. Michael Shermer, *Why People Believe Weird Things: Pseudoscience, Superstition, and Other Confusions of Our Time*, New York: W. H. Freeman and Company, 1997, p. 277.
15. Ibid.
16. Ibid. 275.
17. Park, 'Future Schlock'.
18. Carol Tavris, 'Call Us Unpredictable', *New York Times*, 2 January 1998.
19. Anthony Aveni, *Behind the Crystal Ball: Magic, Science, and the Occult from Antiquity through the New Age*, New York: Times Books, 1996, p. xii.
20. Ibid. 346–7.
21. Ibid. 339.
22. Ibid. 339–40.
23. Roger Altman, 'The Nuke of the 90s', *New York Times Magazine*, 1 March 1998.
24. Robert Kuttner, 'Globalism Bites back', *The American Prospect*, March–April 1998.
25. David Moberg, 'Power Grab', *The Progressive*, March 1998.
26. Lester Thurow, *The Future of Capitalism: How Today's Economic Forces Shape Tomorrow's World*, New York: Penguin, 1996, p. 326.
27. Altman, 'The Nuke of the 90s'.
28. The market not only sells scientific and religious produces but also mixes them together to create glossy products such as 'phone psychics' and makes money out of them. See Stephen Glass, 'Prophets and Losses: The Futures Market for Phone Psychics', *Harper's Magazine*, February 1998.
29. Thurow, *The Future of Capitalism*, p. 303.
30. Ibid. 309.
31. George Soros, 'Toward a Global Open Society', *The Atlantic Monthly*, January 1998.
32. Jerome K. Jerome, *The Idle Thoughts of an Idle Fellow*, New York: Henry Holt and Company, 1913, pp. 195–6.
33. Ibid. 196–7.
34. Quoted in Eric Zorn, 'In the City of God', *Notre Dame Magazine*, Autumn 1996, p. 37.
35. Aveni, *Behind the Crystal Ball*, p. 120.
36. Earl G. Graves, 'Staking a Claim on the Future', *Black Enterprise*, March 1998, p. 11.
37. See Timothy L. Jenkins and Khafra K. Om-Ra-Seti, *Black Futurists in the Information Age*, New York: KMT Publications, 1997.

38. Stuart Seely Sprague, *His Promised Land*, New York and London: W. W. Norton & Company, 1996, pp. 25–7, 39–40, 43, 51, 52, 58, 61.
39. Jimmy Carter, 'Prayer and the Civic Religion', *New York Times*, 24 December 1996.
40. Kevin Sack, 'From Power to a Higher Power: Carter's Journey', *New York Times*, 15 December 1996.
41. Rick Bragg, 'She Opened World to Others; Her World Has Opened, Too', *New York Times*, 12 November 1996.
42. Elaine Sciolino, 'A Painful Road from Vietnam to Forgiveness', *New York Times*, 12 November 1996.
43. Seth Mydans, 'Inside a Wartime Brothel: The Avenger's Story', *New York Times*, 12 November 1996.
44. Jurek Becker, *Jacob the Liar*, New York: Arcade, 1996.
45. Jerome, *The Idle Thoughts of an Idle Fellow*, p. 209.
46. F. Barrows Colton, 'Your New World of Tomorrow', *National Geographic Magazine*, October 1945, pp. 385–410.
47. James Botkin, Dan Dimancescu and Ray Stata, *Global Stakes: The Future of High Technology in America*, Cambridge, MA: Ballinger Publishing Company, p. 163.
48. Stanley Woods, *Western Europe: Technology and the Future*, London: Croom Helm, 1987, pp. 96–7.
49. Ben David quoted in Shiv Visvanathan, *Organizing for Science: The Making of an Industrial Research Laboratory*, Delhi: Oxford University Press, 1985, p. 262.
50. Woods, *Western Europe*, p. 94.
51. Rajni Kothari, *Political Economy of Development*, Poona: Gokhale Institute of Politics and Science, 1971, p. 14.
52. Pradeep Bhargava, 'Political Economy of Regional Cooperation in South Asia', in Bhabani Sen Gupta (ed.), *Regional Cooperation and Development in South Asia*, vol. 2: *Political, Social, Technological and Resource Aspects*, New Delhi: South Asian Publishers, 1986, pp. 157–9.
53. Ibid.
54. Quoted in Felix Moses Edoho, 'Globalization and the New World Order: Toward an Inclusive Framework in the Twenty-First Century', in Felix Moses Edoho (ed.), *Globalization and the New World Order: Promises, Problems, and Prospects for Africa in the Twenty-First Century*. Westport, CT: Praeger, 1997, p. 199.
55. Aijaz Ahmad, 'Nation, Community, Violence', *South Asia Bulletin*, 14(1): 27, 1994.
56. Ibid. 30.
57. For a detailed discussion, see 'The Production of Locality', in Arjun Appadurai, *Modernity at Large: Cultural Dimensions of Globalization*, Minneapolis: University of Minnesota Press, 1996, pp. 178–199.
58. Howard P. Segal, *Future Imperfect: The Mixed Blessings of Technology in America*, Amherst: University of Massachusetts Press, 1994, p. xii.
59. See Kaoru Yamaguchi, 'Establishing a Higher Institution for Future-Oriented Studies', in Kaoru Yamaguchi (ed.), *Sustainable Global Communities in the Information Age: Visions from Futures Studies*, London and Westport, CT: Adamantine and Praeger, 1997.
60. See Dan Watanabe, 'Keiki [children] Can Help Us Navigate on Voyage to Future', *Honolulu Star-Bulletin*, 24 April 1995. If the vision is strong families, and the goal is good health, then one benchmark could be reduction in the level of teen pregnancy by *x* per cent by the year 2000.

61. Arnold Toynbee, *Surviving the Future*, New York and London: Oxford University Press, 1971, p. 153.
62. Ibid. 154.
63. Ibid.
64. Ibid. 152–4.

9
Futures Studies and Future Surprises

GRAHAM H. MAY

A few thousand years ago people thought the Earth was the centre of the universe. A few hundred years ago people thought the Earth was flat. Fifteen minutes ago you thought we were the only beings to inhabit this planet. Who knows what you'll know ten years from now.

'K' (Tommy Lee Jones) in 'Men in Black'

As part of a scenario-planning exercise in the mid-1990s to examine the likely future for the airline, the British Airways Strategy Development Group produced a short list of changes that had occurred in the ten years from 1985 to 1995.[1] These included the fall of the Berlin Wall, the election of Nelson Mandela as President of South Africa and a civil war in Europe in which over 100,000 people were killed. Today these events are all history, and as such, perhaps, not particularly remarkable, but who would have predicted them in the mid-1980s?

In 1985 the East German state showed no sign of collapse; indeed, even in May 1989 on a wet dull day I stood on the Western side of the border between the two Germanys, near Duderstadt in the then West Germany, and saw the double fences, with the free-fire zone between, stretching across the countryside, like a scene from a Cold War spy movie. For me it summarised the division of Europe that had been part of the whole of my adult life. At the time it did not occur to me that a few months later, along with many others, I would be watching in amazement the scenes, flashed around the world on television, of the Wall being breached. Two years later I was walking in East Berlin, something I would never have imagined possible before 1989. German friends confirmed that though they had wanted reunification they had not expected to see it in their lifetimes. By the late 1990s, not only is Germany reunited but the Soviet Union, that evil empire in Western eyes, no longer exists and the Russian Federation is involved in collaboration with NATO, its sworn enemy only a decade ago.

In 1985 Nelson Mandela was a prisoner on Robben Island, as he had been since 1964, and the apartheid regime still ruled South Africa. It seemed most likely that Mandela would die in jail or possibly be released when he had become so ill as to be no longer considered a threat. Ten years on he was President of a government elected by the majority of the population of South Africa and respected by many who had previously regarded him as a dangerous terrorist. The fear in Europe in 1985, although fading a little from the Reagan–Brezhnev era, was of superpower conflict, not civil war. The decline of one threat seems to have allowed other conflicts to resurface. Perhaps we should have foreseen it, but few did.

These are just examples of the surprising changes that occurred in the decade from the mid-1980s to the mid-1990s. Over a longer period there have been even greater shifts, which, looking back on them, may seem to have been inevitable but from the perspective of the time probably seemed most unlikely. History is full of empires that have grown and then declined. At the end of the last century it was claimed that the sun never set on the British Empire, so many areas of the globe were part of it. Few in Great Britain, at least, anticipated the changes that would occur in the next 100 years. In 1997 almost the last remnant of that empire, Hong Kong, was returned to China, and Great Britain, a small off-shore island in Europe, was still having difficulty understanding its post-imperial position in the world.

Two thousand years ago much of Britain itself was part of an empire dominated by the Romans. The Roman Empire, which covered most of Western Europe, lasted for about 500 years; the British Empire for about 250 and the Soviet for little more than 50. While they were at their height few could envisage their end, though today we regard their decline as historical fact and even inevitable. Within their period of history it was difficult to imagine how they might end, because they appeared to control the levers of power. The Romans had military dominance and administrative control. The British navy ruled the seas and the Soviet nuclear arsenal stood toe-to-toe with the Americans. At the time their dominance seemed permanent, but in time, as circumstances changed, they passed into history. The Americans and the West may have 'won' in the conflict with communism, but the nature of that 'victory' was not as expected. It was certainly not as Marx had predicted and as seemed quite possible in the 1960s as socialist revolutions occurred in many parts of the world. At that time the West was almost paranoid in the face of the 'Soviet Threat'.

In the historic town of Williamsburg in Virginia the world of 1774 is re-enacted for the benefit of late-20th-century tourists. One of the roles to be observed is that of an overseer of a tobacco plantation. He talks of slavery as inevitable, the only way to run the plantation, and anticipates that it will last for a thousand years. From the point of view of the 1990s this is shocking and surprising, but in 1774, with the slave trade at its height, it probably seemed quite normal. Who then could have predicted the changes, technological, economic, social and political, that brought the production processes based on slavery to an end? Perhaps every generation finds it difficult to imagine that its own situation is likely to be as temporary as those that have gone before, despite the evidence of history on which we can draw. It has even, of course, been suggested that we have come to the 'end of history',[2] as the West and Western culture become ever more dominant. The future, therefore, is predictable; it will continue on the current trajectory.

So it might, but a surprise-free future would be, in the light of history, the most surprising outcome. It is, in fact, out of this realisation, that the future is not just a continuation of the past, that futures studies has developed. It was, for example, the failure of traditional forecasting methods in the 1970s that encouraged Shell to develop alternative scenarios of the future as the basis for its strategic planning. But even in the range of assumptions it has made about the price of oil, a basic aspect of its operation, it has not always included the eventual outcome. Perhaps the aim of Western professional futurists is to maintain the dominance of their corporate clients, but there is no guarantee that they will succeed. Many companies, for example, have failed to foresee the growing concern for the environment and have underestimated the impact this would have on their operations. Shell itself, although the scenario planners did include the eventuality in their work, did not take the threat of consumer boycotts seriously; but following the publicity gained by Greenpeace in opposing the disposal of the Brent Spar oil platform it was forced to backtrack, eventually even inviting Greenpeace to help it decide what to do with the platform. Such companies, which are, if anything, the main force for Western economic imperialism, are far from infallible. It has, for example, been shown that relatively few of the major companies of the 19th century still exist today,[3] and many originally Western industries have been overtaken by Asian, particularly Japanese firms. Many of these firms have now invested in Western countries, particularly the USA and Britain, bringing with them very different management cultures. The influence is far from one-way, and as communication becomes easier it is likely to become less and less so. Indeed, it has been suggested that the culture within which some Asian companies have developed gives them a competitive advantage over the West. A Western game, perhaps, but as the English cricket and soccer teams have found others now play the game better than the originators. Then, just as conventional wisdom seemed to have accepted the coming Pacific Century, came the events of late 1997.

It is sometimes suggested that the control of communications technology provides a means of cultural imperialism and domination by American culture and the English language. This has some credibility, but contrary examples do exist. The Iranian Revolution, in which Ayatollah Khomeni overthrew the Shah, who was attempting to Westernise the country, was made easier by the cassette tape-recorder. Even the Shah's secret police were unable to intercept the cassettes on which the Ayatollah's recorded speeches were circulated. The result appears to have been a positive rejection of Western values and culture and a re-assertion of traditional attitudes. The latest technology that it is sometimes said will increase Western dominance is the Internet. Perhaps it will, but it should be remembered that the technology originated as a system for the American military that was designed to withstand nuclear attack by re-routing around breaks in the network. It is difficult to believe that the designers of the system imagined either the number of uses to which the Internet is now being put or the anarchy that makes it almost impossible to control. One of the major concerns of society and particularly authority is the uses, be they political or pornographic, to which the Net is now being put.

One of the features of the so-called modern Western world is control, the belief that it is necessary only to know enough about how things work for the ability to control future events to follow. From this perspective the world is essentially simple and can be understood as a series of simple systems. In some areas of the natural sciences this provides sufficient certainty to allow accurate prediction, but social, economic and political affairs are proving more complex, resisting attempts to predict and control them. Daniel Bell's contention that systems analysis would soon allow totally rational decisions has still to be achieved and there are increasing doubts that it ever will.[4] This may be in part because, as Casti argues, simple systems have:

- predictable behaviour;
- few interactions and feedback/feedforward loops;
- centralised decision-making; and
- decomposability, that is, they can be understood by reducing them to their component parts.[5]

On the other hand, the more recent concepts of chaos and complexity theory—often termed 'postmodern', and brought together in the term 'chaotics'[6]—suggest that the systems in which we live are not simple but complex. Complex systems are:

- unpredictable, often leading to counterintuitive results;
- full of feedback/feedforward loops, which enables them to respond to and absorb shocks;
- decentralised with many decision-making points; and
- irreducible, in that slicing them up into subsystems destroys their very nature.

In such circumstances diversity has positive value; domination by one culture would be counterproductive and, just as in the case of monoculture in agriculture, positively dangerous to the survival of the human species. Perhaps the Internet is one of the developments that will help us avoid that particular trap?

This perhaps goes some way towards explaining Naisbitt's 'Global Paradox',[7] that 'The larger the system, the smaller and more powerful and important the parts.' This is reflected, he argues, in the apparently contradictory developments of a global economy, which suggests a concentration of power in the hands of major transnational companies, and both a breaking-up of these giants into smaller parts and a growing cultural and political tribalism that distributes power more locally. We may be moving towards a global economy but culturally and politically there are trends in the opposite direction, as past mega-states like the Soviet Union break up into their constituent parts. Almost as a reaction to the apparent domination by Western companies, local culture is reasserting itself. 'On the face of it, the Asian tigers can appear palpably Western: it is only when you peer a little more closely that you realise their cultures are a bewildering synthesis of the Western and the local.'[8] The probable emergence of China as the largest economy in the world and India as the most populous country is likely to have a major impact, particularly when it is related to the Chinese and Indian communities in the rest of the world, which continue to maintain their cultural identity. Even in the West itself, where groups had apparently been assimilated into countries like the United Kingdom, they are now pressing for greater autonomy and distinctiveness as Scots, Welsh or Cornish. Even after several centuries of assimilation and attempts to eradicate local languages and culture they are again increasing in significance.

There is here another paradox, that as English is said to be becoming the international language native English-speakers are at both an advantage and a disadvantage at the same time. Native speakers of another tongue may learn English to communicate internationally, but retain their own language for 'private' conversation. Native English-speakers have no such privilege; neither do they have a clear international language to learn and in consequence become lazy, assuming everyone else will speak English, nor do they retain their own 'private' means of communication. It can be an advantage to be a native English-speaker, but there is a downside too as the language becomes internationalised.

It is into this increasingly complex, interrelated and paradoxical world that futures studies is developing. It may have originated in Europe and America, in industry and the military, but as an academic activity it is still in its infancy. Many of those currently involved are white, Anglo-Saxon, middle-aged males, the author included, but it is clear that there are many other forms of development that go far beyond the Western business bias of 'official' futures. Slaughter, for example, indicates that the futures field ranges from business-dominated futures research, through futures studies to futures movements, which are concerned with stimulating, reconceptualising and leading change and include alternative lifestyles and transpersonal psychology.[9] One of the most significant forces, mainly, but not exclusively, in the West, in the last thirty years has been the environmental movement. It has brought about fairly major shifts in attitudes, such that the environment is now one of the main causes of political action.

As an academic activity there are as yet few futures studies courses, and there are indications that they are developing as quickly outside the West as within it. Those that do exist, such as the Masters programmes at Houston, which has existed for over twenty years, and Hawaii in the USA, and the course in Leeds in the UK, launched in 1996, make conscious attempts to present a multicultural perspective. Elsewhere programmes or research institutes either exist or are under development in Pakistan, Korea and China, among others. While the World Future Society is predominantly North American, the World Futures Studies Federation is more international and multicultural, and the Future Generation Alliance Foundation, based in Kyoto, Japan has a Buddhist base, which is fundamental to its approach. In many parts of the world a wide range of community-based initiatives devoted to social innovation, new economics, environmental activism and alternative lifestyles are concerned to create a preferable future rather than predict the continuation of business as usual.

Futures studies is not yet a discipline in the accepted sense of the term. This is partly because it has not yet established an agreed corpus of knowledge or methodology, but it is also due to the very nature of futures studies itself. Western academic and scientific endeavour has been based on empiricism, the search for knowledge through the establishment of facts. As de Jouvenel pointed out, futures deals not in 'facta' but in 'futura', because provable facts belong to the past.[10] Empirical knowledge has its place in futures but it is insufficient; other forms of knowledge, such as belief, faith, ethical and moral conviction and the envisioning of future images become equally important in creating the future rather than understanding the past. Western culture is frequently seen as lacking such characteristics.

Developing an Argument for Futures Studies

'If anything is important it is the future. The past is gone, and the present exists only as a fleeting moment. Everything that we think and do from this moment on can only affect the future. and it is in the future that we shall spend the rest of our lives.'[11] History is important. It has many lessons to teach and failure to heed them is one of the easiest ways to make mistakes in the future.[12] History also carries with it many experiences both personal and societal that influence attitudes to the present and the future. In many situations these attitudes, which are derived from the past, determine our approach to the future. To create change we may need to break away from this domination by the past and think anew. This is often difficult, but if we are to solve many of the world's problems, it is necessary.

One way to assist this may be to examine the way in which we refer to different periods of time. The English language is not alone in using the words 'past', 'present' and 'future', each suggesting the singular. There was one past—history rather than histories; there is one present, not a series of presents; and there will be one future. If, however, we accept the existence of effective human choice, there must be the potential for different *futures* to occur. Easy as it is to explain, when we are used to thinking about *the past* and *the present*, it can be still be difficult to accept that *futures* are multiple. It seems to make the future different.

But is it as different as all that? Consider the experience of an Englishman in Boston, Massachusetts. Some of the events that took place in and around Boston in the late 18th century are probably familiar to most English schoolchildren, but seeing them from an American perspective casts them in a new light. The British version of history generally regards the 'Brits' as the good guys, fighting bravely for right. Following the 'Freedom Trail' through Boston poses some interesting questions. The American view is different. We, the Brits, were the baddies!

Although the past has happened and left evidence and memories, difficulties remain. The interpretation of the events can be very different when viewed from different perspectives. Which, if any, is right? Some of the most intractable political problems of today, such as those in Northern Ireland, the Middle East and the former Yugoslavia, originate in very different perspectives on history. Perhaps the main contribution that futurists can make to the resolution of these conflicts is to emphasise the existence of, and need to understand *histories*, in the plural; that, although there may strictly have been one past, different human perceptions effectively create multiple pasts that are often incompatible. Only by recognising the significance of these pasts can the varying presents be understood, with a view to achieving a future that will go some way towards meeting the conflicting preferences.

The difficulty of being certain about the past is also revealed in the investigations that follow major disasters. Despite all the evidence it is possible to collect there often remains an element of doubt about the causes. The event occurred, but why? Can blame be attributed? A rail crash may have been directly due the driver passing a red light, but if cab radios or automatic train protection had been installed it could have been averted. It is difficult to be categorical in attributing blame. The driver was at fault, but relatively simple safety precautions that could have prevented the accident had not been installed because others decided not to do so. Were they not also responsible? The reliance on juries at criminal trials is another example of the difficulties of being certain about the past. The jury system is based in the belief that if a group of different individuals are convinced about what happened the chances of getting to the truth are greater. There remains an element of doubt, as some notorious miscarriages of justice have shown, but the system is defended for much the same reason as Churchill is said to have favoured democracy: 'Democracy is the worst form of government, apart from all the rest.' Accuracy is not assured but inaccuracy is most probably reduced.

The present is not that different either. In a sense there is one present, but each of us has only a partial experience of it, depending on where we live, what information we receive and how we interpret it. My present is not the same as yours, and even if we are in the same place at the same time our perceptions may be very different. An interesting example occurred on a flight from Manchester, in the UK, to Atlanta, Georgia. A small card headed 'Health Alert Notice' was handed to all passengers. A note to the physician read, 'The patient presenting this card has recently been abroad . . .'. From a US perspective that was true, but from a British one it was not. The Americans had been abroad, the British had been at home. The British were going abroad, not coming from! Both were real presents—they even related to the same event—but the experience was different. As individuals we may experience only one, but the existence of different perspectives and experiences effectively creates a number of presents occurring simultaneously rather than just one. A result of global communication, particularly television, is that the different sides to a dispute can now be heard and seen. We may not agree with them but unlike earlier conflicts the alternative perspective of the other side can now be heard.

If it can be shown that a better understanding of the human condition can be gained by thinking of multiple pasts and presents, the idea of alternative futures becomes less disturbing. Alternative pasts and presents make the notion of alternative futures unexceptional and easier to accept.

Thinking seriously about the future before it happens is not a new concept. It is enshrined in common sense in the English language in phrases such as 'Look before you leap', 'Forewarned is forearmed', and 'Prevention is better than cure'. All suggest that the value of thinking ahead about potential future events has been understood for a considerable time. Other languages probably have similar concepts.

Although our relationship to the future cannot be proven absolutely, it is reasonable, on the evidence of history, to conclude that history is made, not given, and that human action has an impact, sometimes a determining one, on the shape of events. We are the future of the past; our lives are heavily influenced by the actions of previous generations. Equally, we are the past of the future; our actions will influence the potentials of those who follow us. This has always been true but there is increasing evidence to suggest that the impact of current generations on the future is greater than ever before. As the Italian industrialist Aurelio Pecci, founder of the Club of Rome, argued, 'The future will no longer be a mere continuation of the present, but a direct consequence of it.'[13] We are beginning to realise that what we do could even affect the chances of the continuation of human and other life-forms on the planet.

This re-emphasises our responsibility to future generations, because just as we inherited the world from our ancestors so future generations will inherit it from us. Non-Western cultures that were less taken up by the attractions of technologically based industrial progress have important lessons to teach in a world in which we are not owners but merely tenants with a relatively short lease. The concept of 'sustainable development', which is defined as 'Development that meets the needs of the present without compromising the ability of future generations to meet their needs,[14] is essentially based on this idea. At face value the definition is reasonable, but it raises many issues, not least because it is clear that we are not meeting even the basic needs of many in the current generation. If we are unable or unwilling to do that, what right have we to pretend to act in the interests of future generations, even if we were able to know what they will be?

The definition does at least provide, if we wish to take it seriously, the necessary first step in making decisions for the future. It provides, however fuzzy, an image of a different future to work towards. Without such an image of a desirable future there is no reason to decide to work towards a particular direction, because 'If you do not know where you are going, any road will take you there.'[15] It is in is this way, as Popper argued,[16] that the future pulls us as much as the past pushes us.

Change is a natural and ever-present process. The passing of the seasons, life and death, new experiences, knowledge and ways of doing things, continue from day to day and year to year. In such circumstances new challenges and issues are always likely to arise. Some situations, such as accidents or natural disasters, force us to react, but even they can often be anticipated, if not exactly predicted, and precautions taken to reduce their impact and preparatory measures put in place to enable a quick and effective response. In other situations anticipation, thinking about possible events before they happen, can provide time to review a wider range of options and avoid the restrictions imposed by crisis management. Techniques such as impact assessment, risk assessment and foresight make more informed decisions possible. Waiting for events also reduces the possibility of positive influence over the future, of creating change rather than just being subjected to it, and making the future happen for us rather than to us.

Thinking about the future is easy, we do it all the time. Preparing a meal, planting crops and planning a journey all require thinking about the future. The difficulty occurs when the future we prepared or planned for is not as we anticipated it would be. Experience suggests that this mismatch occurs enough to make us uncomfortable and to encourage us to doubt the value of thinking about the future. We all have to live in the present and coping with the present is often enough; it may leave no time for thinking about things that may never happen. These present concerns are also more real and demand attention, whereas consideration of potential problems may be seen as a waste of resources and can be put off until they become more pressing, even though 'a stitch in time may save nine'.

Thinking about the future is also complicated by the different kinds of future thoughts that we can have, which often get confused. There may be a wide range of possible futures that we believe could happen. At one extreme are our hopes, those situations we would find favourable, and at the other our fears, those futures that would be unpleasant for us. Depending on whether we tend to be optimistic or pessimistic, our view of the likely future will be influenced towards one extreme or the other. Equally important are our attitudes to determinism and choice. Do we believe that the future is determined by preordination, the inexorable march of history, genetics, our environment or the economic and social structure of society? If so, only prediction is possible, we cannot influence the future and thinking about the future has little value. Choice, on the other hand, may introduce too many possibilities for us to deal with, because the future becomes conditional and unpredictable

Our experience of forecasting is also not encouraging. Wise suggested that only 40 per cent of forecasts of technological developments between 1890 and 1940 had been successful by 1974 and that forecasting human affairs was even more inaccurate.[17] Sherden, having examined a range of forecasting situations, has suggested our performance is little better than chance.[18] The record does, however, provide a number of lessons as to why this may be the case. Forecasts are often too narrow, because in being made, for example, by economists they ignore the non-economic issues that are often significant. They also frequently suffer from overoptimism, timidity or the assumption of freedom from surprise. It is also important to examine the purpose of forecasting. It is usually assumed that forecasts are made in order to foresee the future, but as Pohl points out,[19] if they are accurate forecasts are of little use, because there is nothing we can do to influence the future. In fact forecasts are often made for other purposes, such as to warn of potential problems, assist choice between alternative courses of action or, less desirably, for propaganda.

It is often thought that whereas we can have knowledge of the past and present there can be no knowledge of the future, as it remains totally uncertain. Further examination of the nature of knowledge reveals that the past and present may also be uncertain as we search for further information about history and are faced with contrary interpretations of past and present events. Meanwhile certain things about the future, such as sunrise tomorrow and the date of the next American Presidential election, can be predicted with reasonable certainty. Uncertainty is not the sole preserve of the future, it is inherent in the human condition, past, present and future. The future is consequently not so different.

There is the danger of colonising the future and closing off options by projecting currently dominant ideas and values into the future and assuming they will continue to dominate; but as we have seen at the start of this chapter, there is ample evidence from history to suggest that this is far from inevitable. Present dominant cultures and values may continue, but one of the main arguments for futures studies is to encourage careful consideration of the potentials of the future rather than blunder forward into a future that without thinking about it seems inevitable.

Everybody has an interest in the future as the territory in which they and their descendants will spend the rest of their lives. The future is therefore too important to be left to the professional futurists. The development of futures studies, particularly within the educational context, is concerned with encouraging 'students to understand the ways in which they can help to shape their own futures by offering them the skills to participate in that shaping'.[20] The futures movement also includes a wide range of activists working with communities to achieve similar empowerment in other parts of society. As has been suggested elsewhere,[21] there is considerable evidence to conclude that 'The Future is Ours' if as a human race we wish to take up the challenge. The development of futures studies has an important role to play in helping us face the issues involved.

A Learning Approach to the Future

Experience of dealing with the uncertainty of the future can lead us to dismiss the idea of attempting to prepare for it because it is just too unpredictable to make the effort worthwhile. This is quite understandable, but it ignores the other side of our paradoxical relationship with the future, that we know what we do is likely to have an influence on what happens. Indeed, it is possible to argue that, although it has always been true that one generation plays a significant role in creating the circumstances in which those that follow will live, the implications of this have only become so apparent in the late 20th century. We are now not only capable of significantly altering the life chances of succeeding generations, but have also become aware of that capability while knowing that we are not able to predict with certainty what the results of our actions will be. One way of dealing with this situation is to adopt a learning approach to the future.

Approaching the future as a learning process suggests a number of necessary characteristics, including:

- Open-mindedness;
- Diversity;
- Flexibility;
- Creativity;
- Accepting there will be few final solutions;
- Continually seeking information;
- Giving up control to maintain influence;
- Welcoming ideas invented elsewhere;
- Accepting a participatory approach; and
- Regarding strategies as learning devices.

Open-mindedness

Because many situations involve uncertainty it is necessary to be open to different perspectives based on varying values, positions and views; to accept the existence of differences and even conflict between different perspectives and attempt to work with the tensions and encourage debate. This is often not easy. All sorts of justifications are used to support 'our way' as the one true path: religion, political ideology, nationalism, race, tradition, academic discipline etc. We like to think that we are right. You, if you disagree with me, are, to put it politely, misguided. They are clearly wrong, after their own ends, evil, corrupt or generally up to no good.

Accepting the validity of differing perspectives challenges many of our preconceptions. It may imply that simple notions of right versus wrong need to be replaced by a realisation of more complex interrelationships of rights and wrongs, depending on how the issue is approached. The adversarial traditions of politics, industrial relations and the law may need reconsidering. In the UK the layout of the House of Commons enshrines the adversarial tradition with government and opposition facing each other across a divide two sword lengths apart. It is fine for hurling insults at each other but does not encourage considered debate of the complex issues of modern government. It also relates to a world long past; few Members of Parliament wear swords these days, and if they wished to inflict harm on each other much more effective ways of doing so are now available. A learning approach to the future would recognise that we do not always have the right solution, or may not even be asking the right question; that lessons can always be learned from others and that bringing different perspectives, which are often more complex than either/or, to bear on an issue is likely to be an important part of finding a way forward.

Diversity

Recognising that cultural diversity provides a valuable resource in human affairs may be as important to the survival of humankind as biodiversity is to the ecosystem. The case for biodiversity is based in the susceptibility of single strains to devastation by disease or pests and the potential of as yet undiscovered species. The loss of cultural diversity could have similar risks for the survival of humanity. Loss of variety such as the loss or eradication of local languages in the face of economic globalisation or political persecution, Monbiot suggests,[22] results in not only the loss of cultural identity and diversity but the growth of humankind's fastest-growing disease, anomie; the loss of those cultural reference points that make life worth living. Groups that have lost their language and culture and are not integrated into society, he contends, are prone to alcoholism, drug abuse and suicide.

Diversity means complexity, but, as Monbiot argues, although 'living with complexity is a messy, difficult business, which requires constant responsiveness, creativity and goodwill' (a learning approach, in other words), 'simplicity is very much harder', because it limits the perspectives available and does not reflect reality. It is becoming increasingly clear that many issues are like this, and that simplifying them only produces unexpected and unfavourable results.

Flexibility

In the face of complexity and uncertainty it is necessary to remain flexible, to be prepared to reassess situations and plans as circumstances change. One of the most difficult features of this is to make decisions about when to be flexible and when to hold to established positions. One of the most moving sequences of the musical 'Fiddler on the Roof' is the song 'Tradition', which clearly exposes the problems of societies in adapting to change. Tradition and established ways of doing things are powerful and valuable forces that provide a focus for human interaction, but change questions them, and it can become disadvantageous to retain them beyond their usefulness. For example, if all the old traditional environments had been maintained many of the things we now value would not have been developed. If citizen groups had been as powerful in the last century as they are now, many of the Victorian buildings we now wish to preserve would never have been built. Many of them were built in open countryside or replaced older buildings that would now be conserved.

Changing circumstances call for changing responses. As Kelly points out, perfectly pure rules that are applied absolutely are not resilient to competing strategies.[23] They eventually lead to death rather than sustain life because they fail to adapt to new challenges. Viruses evolve new strains to counter the drugs used to protect us from them, so medical science cannot stand still. Human society is similar; established ways of doing things gradually become less effective as new challenges arise. To remain effective it is necessary to balance the need for retaining links to the past with the need to develop continually evolving responses to new concerns, and balance is a dynamic concept.

Creativity

Being creative and responding to change and uncertainty require a willingness to accept that not all decisions and actions will be correct. Automatically punishing decisions that turn out to be mistaken discourages the kind of innovation that is necessary to deal with new situations in which uncertainty is present. Both Kelly and Toffler note that mistakes and imperfections play an important part in creating long-term sustainability and keeping systems moving forward.[24] Unfortunately, in many organisations making mistakes is regarded as something to be avoided because it wastes precious resources, whereas, in the longer term, not allowing mistakes and learning from them is probably costlier because it stifles the creativity needed to deal with evolving situations.

Jungk and Mullert suggest that creativity requires a preparedness to

- think the otherwise unthinkable;
- be enterprising and inquisitive;
- be non-conformist and flexible;
- be open-minded to the irrational and off-beat;
- take a chance on being wrong or failing;
- shun cynical, know-all and perfectionist attitudes; and
- stand up for cranky ideas.[25]

These are not normally the expected characteristics for success in employment or academe.

Few final solutions

Accepting that there are few final solutions may mean that some problems cannot be 'solved' and taken off the agenda, but will require continuing action as they evolve. Despite experience of solutions to problems that do not work out as expected, do not completely solve the original problem or even create new ones, the search for the quick fix remains popular. Believing that problems can be ticked off rather than needing continuing care is more attractive but frequently less effective. Such situations require continual monitoring and management rather than a grand plan that is followed through to completion. The future seldom stands still long enough to be implemented as we once thought it would be. Modifications are needed as experience is gained and unexpected reactions or new influences emerge. Unfortunately this often runs counter to the preference for strong leadership and is frequently derided as 'policy-making on the hoof' rather than sensible adaptation to changing circumstances.

Information

Information plays a critical role, particularly in evaluating actions and closely monitoring developments in areas with significant impacts on our interests. These are not always obvious; Schwartz, for example, advises scenario researchers 'To look at the world as horses do; because new knowledge develops at the fringes'.[26] This can be difficult, because we are usually selective in what we notice and seek out. Academic knowledge, for example, is divided up into disciplines, which focus on a particular area of activity. This tends to marginalise developments outside the boundaries, which, experience of forecasting shows, are often critical in invalidating predictions. The reductionist approach of traditional scientific method is prone to similar shortcomings, encouraging the collection only of information that is considered relevant to the initial hypothesis while disregarding other information that may prove more significant. New information, while important, is not always reliable or comparable with existing knowledge. Consequently it requires judgement to decide whether it represents only short-term fluctuations of little significance or the onset of major change. Failure to search for the new and continually assess its importance as it emerges, on the other hand, restricts the information available. As a result, potential problems are not identified in their formative stage but are only recognised when they have become established and probably more difficult to deal with.

Giving up control to maintain influence

Openness, flexibility, creativity and scanning for change require an acceptance of limited control. In a multi-player game with a complex system of interactions the actions taken may not have the effects anticipated if other players respond in unexpected ways. One of the institutions that is finding it hard to adapt to this feature of our future is the nation-state. Used to the idea of sovereignty, a very special degree of control, particularly over the people within their own territory, governments are finding it difficult to accept that the best way of retaining influence may be to give up control over areas they believe to be sacrosanct. The globalisation of the economy means that a single national government can no longer control the value of its currency or the transnational firms that dominate the market in many areas. Both satellite broadcasting and the Internet evade their attempts to control the information available to their populations, and environmental pollution takes no cognisance of national boundaries. An effective response may involve some power being surrendered to larger groupings, such as the European Union or the

United Nations, and paradoxically some to smaller units of more local control where the needs of local communities can be more easily identified.

Welcoming ideas invented elsewhere

We always like to think that our ideas and ways of doing things are best. Opinions about which is the most democratic system of government, or the best environmental planning system, will frequently depend more on whether the assessor is American, British, French, German or other nationality than on the features of the systems themselves. Having pride in one's own country or our own achievements is no bad thing, but if it blinds us to the benefits of learning from others and reassessing our own ways it becomes a serious disadvantage. Not-invented-here is a severe handicap in a flexible approach to the future and some of those lessons may be best learned in the most unexpected places. Modern communications and transport provide the opportunity to look more widely than ever before for new ideas, but to realise the opportunity requires some of our psychological inhibitions to be overcome.

A participatory approach

Everyone has an interest in the future. It is where we will all spend the rest of our lives. In the past, futures has often been regarded as esoteric and elitist, dominated by big business, government and a few academics. Perhaps it has been so because such groups wished to retain their power over the future and make it happen to suit them. Experts are necessary. Without expert knowledge few of the things we take for granted in modern life would be possible, but there is also a growing demand for self-expression and self-determination. Ironically, it arises, at least in part, because as more individuals become educated and expert in particular fields, the limitations of expertise are more clearly understood. Even experts make mistakes, have limited knowledge, and in some cases set out to mislead. Consequently we trust them less, want to know why we should follow their advice, and decide for ourselves whether to accept it or not. There is at least the possibility that developing information technologies will provide the opportunity to do this.

Learning strategies

It may be assumed that the acceptance of change and flexibility obviates the need for strategy and planning. Far from it. Without visions of the kind of future we want, there is no reason for making decisions, because anything will do. Strategy is necessary, but a strategy that is itself structured as a learning process. As Pedler *et al.* argue, plans need to be developed, formulated and revised as situations develop.[27] Decisions should be seen as conscious experiments, rather than set solutions. 'Deliberate small-scale experiments and feed back loops are built into the planning process to enable continuous improvement in the light of experience.' This emphasises the crucial roles of learning and judgement. When is it best to stick with the plan and when to change in the light of evolving circumstances? That itself is a skill to be learned.

An Exciting and Threatening Future

Regarding the future as dynamic and uncertain can be unnerving and even threatening. There are enough reasons in the modern world for thinking it so. But it is also exciting and liberating. There is always something new to be created or discovered. There are no experts who know it all. The impertinent question may have more value than the pertinent answer. Learning for the future becomes less a matter of learning an established set of fixed knowledge to apply to known problems—important though it is to understand how we got where we are—and more a search for the new and the continuing development of capabilities to create our futures and deal with the surprises that occur.

Notes

1. K. Moyer, 'How to Structure Scenarios', paper to the Conference 'Practically Implementing Scenario Planning', IIR, May 1997.
2. F. Fukuyama, *The End of History and the Last Man* London: Hamish Hamilton, 1992.
3. A. de Guies, 'Companies: What Are They', *RSA Journal*, June 1995: 26–35.
4. D. Bell, *The Coming of Post-Industrial Society: A Venture in Social Forecasting*, London: Penguin, 1973.
5. J. L. Casti, *Complexification: Explaining a Paradoxical World Through the Science of Surprise*, New York: HarperCollins, 1994.
6. G. Anderla, A. Dunning and S. Forge, *Chaotics: An Agenda for Business and Society in the Twenty-First Century*, London and Westport, CT: Adamantine and Praeger, 1997.
7. J. Naisbitt, *Global Paradox*, New York: Morrow, 1994.
8. M. Jacques, 'Sleeping Giant Awakes to Claim the New Century', *The Observer*, 15 June 1997, pp. 8–9.
9. R. Slaughter and H. Beare, *Education for the Twenty-First Century*, London and New York: Routledge, 1993.
10. B. de Jouvenel, *The Art of Conjecture*, London: Weidenfeld and Nicolson, 1964.
11. E. Cornish, *The Study of the Future: An Introduction to the Art and Science of Understanding and Shaping Tomorrow's World*, Washington, DC: World Future Society, 1977.
12. E. Hobsbawm, 'To See the Future Look at the Past', *Guardian*, 7 June 1997, p. 21.
13. A. Pecci, *One Hundred Pages for the Future*, Oxford: Pergamon, 1981.
14. World Commission on Environment and Development, *Our Common Future*, Oxford: Oxford University Press, 1987.
15. S. Enzer and R. Wurzburger, 'LA 200 + 20: Some Alternative Futures for Los Angeles

2001', executive summary, Los Angeles: University of Southern California, Center for Futures Research, 1982.
16. K. Popper, 'The Allure of the Open Future', *Guardian*, 29 August 1988, p. 8.
17. G. Wise, 'The Accuracy of Technological Forecasts', *Futures*, 8(5): 411–19, 1976.
18. W. A. Sherden, *The Fortune Sellers: The Big Business of Buying and Selling Predictions*, New York: Wiley, 1998.
19. F. Pohl, 'The Uses of the Future', *The Futurist*, 27(2): 9–12, 1993.
20. C. Hooker, 'Future Studies: Its Role in Education', in R. Slaughter (ed.), *Studying the Future: An Introductory Reader*, Victoria: Commission for the Future and the Australian Bicentennial Authority, 1989, pp. 47–50.
21. G. May, *The Future Is Ours: Foreseeing, Managing and Creating the Future*, London and Westport, CT: Adamantine and Praeger, 1996.
22. G. Monbiot, 'Global Villagers Speak with Forked Tongues', *Guardian*, 24 August 1995, p. 13.
23. K. Kelly, *Out of Control: The New Biology of Machines*, London: Fourth Estate, 1994.
24. Ibid.; A. Toffler, *Powershift: Knowledge, Wealth and Violence at the Edge of the 21st Century*, New York: Bantam Press, 1990.
25. R. Jungk and N. Mullert, *Futures Workshops: How to Create Desirable Futures*, London: Institute for Social Inventions, 1987.
26. P. Schwartz, *The Art of the Long View: Scenario Planning—Protecting Your Company against an Uncertain World*, London: Century Business, 1991.
27. M. Pedler, J. Burgoyne and T. Boydell, *The Learning Company: A Strategy for Sustainable Development*, New York: McGraw Hill, 1991.

10
Futures Studies and Foresight

TED FULLER

Futures studies, as a discipline, is colonising the future of futures. The idea that anticipating the future, or being cognisant of the meaning of our actions for future generations, or being able to think in 'alternative' ways is somehow a 'discipline' is a mistake. It is similar to assuming that science is separate from humanity. If the future of futures is successfully colonised by futures studies, however eclectic its philosophies, then foresight will fail to become a general skill or competence. It will become a fragmented specialism whose fallibility cannot be adequately critiqued by those whose experience is essential to understanding.

To 'look ahead' is a natural and ubiquitous activity. Foresight is the skill of making meaning of looking ahead. It is an intuitive philosophical exercise operating in real time and in reality. It is led by, and may lead to, philosophical progress. Futures studies is colonised by notions of progress, and may even be synonymous with the meaning of progress. A worldview of futures studies requires a worldview of 'progress'.

Futures Futures

Studies of the future occur in many domains and in many senses. The domains of futures studies as seen by their proponents are well documented, for example in *The Knowledge Base of Futures Studies.*[1] No exhaustive list of the domains in which an appreciation of the future is undertaken is possible, because looking ahead is a ubiquitous activity in every domain. The issue of plurality in thinking about the future barely need arise, such is the natural plurality of foresight. The issue is the communicated meaning or reasoning of these plural views and the way in which they shape the emergence of aggregate actions.

What follows is a brief visit to some of these domains and the paths that they appear to be on. In each domain, futures studies can be seen as a particular question or issue, not as a separate discipline. These domains are philosophy, culture, alternative thinking, worldviews, the guardians of future generations, forecasting, planning and the futures business.

Philosophical Futures

If futures studies is a way of thinking (about the future), then it belongs in the domain of philosophy. Futures studies is dependent on epistemologies, on principles of what can be known and on philosophical principles of how we can know that. Can we know future epistemologies? I think not. Epistemologies lend us our subjective meanings of our world, and therefore we cannot accurately reflect future meanings without the epistemological perspective of that future time. There are only current ways of thinking about the future. For as soon as we create an epistemological change, it exists. The discipline of futures studies, if so directed, can create epistemological progress, especially if theorising is critiqued for its meaning and value with regard to prediction. *In this domain, futures studies becomes the question 'So what does this or that philosophical principle do to help us contemplate what is yet to be experienced?'*

If social reflexes on Western futures—that is, what people accept as legitimate discourse on the future—are foreclosed, then the foreclosure is paradigmatic, not technological. Western analysis is largely empiricist. How can you talk about anything without evidence? Well the problem with evidence is that it is partial and historical. Evidence should not be ignored. It should be used to inform our thoughts about the future. It should not be relied upon. Evidence is always partial. The world is whole, the future is whole. Evidence is how humans give a pattern to something they can measure. Some high priests of futures are most guilty of factoid predictions. 'Trends' is the fatal word, despite widespread use by futurists, including myself.[2] Empirically 'proven' trends are powerful communications of apparent meaning. But they are not communicating the meaning of their reality, only a pattern made through measurement of the measurable and their 'fit' with existing knowledge—the 'selection of the fittest'. The ontological enterprise of futures studies is interpretativist and its future will remain as interpretativist—making sense of the world, with an objective of relating this sense to a sense of the future.

The future is fallible; that is, our ideas and expectations about the future are fallible. This is the epistemological turn, already made in postmodern or post-Enlightenment philosophies. It is a turn away from belief in scientific objectivity and progress per se from 'scientific' objectivity. Understanding fallibility and examining universal claims from this perspective demands plurality of perspectives. It does not underpin relativism; it does challenge a Western belief that we have an adequate understanding of reality. The concept of fallibility assumes that we produce new realities through the making of new meanings. New realities are new real futures. This is a core concept to underpin the enterprise of futures studies.

Political, Economic, Social and Cultural Futures

The sense in which the future is made by the way in which people conceptually structure their experience is a powerful feature of futures studies. It is in this sense that terms such as 'colonisation' assume meaning. These (PESC) analytiques, with their roots in modernism, are often the metaphors for discourse about the future. They form the framework in which the enterprises of science and technology are given meaning and shaped. Futures studies as a discipline is frequently located, in the West at least, in these domains. As such, futures studies is an amplifier of the status quo, a widening of a narrow path of discovery into a broad highway stretching into the future, but whose direction and boundaries are systemically bounded.

This is not the place to discuss the alternative futures of any of these domains, merely to point out that futures studies as a tool for projecting these domains will not change their fundamental role as particular conceptions of social reality. *In this domain, futures studies becomes the question 'How can the "theories" and meanings constructed within this domain more closely explain observations and guide the understanding and the enacting of progress?'* In this sense, the future is colonised, entrapped in its own discourse.

Futures Studies as 'Alternative Thinking and Critique'

Why should futures studies be about alternative thinking? What has alternative thinking to do with the future? If the values that underlie our relationship with the future are for 'progress', then should we not be concerned with thinking *better*; thinking with better epistemological principles? Are 'alternatives' the same as better? Do we know what 'better' thinking is? Can we know what better thinking is? Are futures studies relative in principle? Can cultural relativism be embraced as cognitive relativism? Is one answer just as good as another?

To some extent the argument for a continued critiquing of the 'status quo' lies in the domain of philosophical development, but has become the domain of a range of 'think-tanks', whose objectives are explicitly about bringing change motivated by a particular mission, such as social change. The UK organisation DEMOS is an example of this.[3] Situating alternative thinking 'in the future' implies that changes are required and that these take time to bring about. These are not, I would argue, futures studies per se. They utilise principles of foresight[4] in the way that logical argument is constructed. This form of 'alternative' thinking brings to light the possible alternatives of different actions and value systems. The presentation of these utilises many of the techniques of futures studies, including impact analysis, simulations and scenarios. *In this domain, futures studies becomes an assertion of preferred futures or a warning of impending futures.* In this domain, futures thinking is a 'subversive' (i.e. relative) influence on mainstream thinking. The effectiveness of such an enterprise is limited by epistemological constraints and by ontological constraints as described earlier.

What futures are possible for subversive think-tanks? Pluralism and the embrace of diversity can, with the prospect of choice, turn subversives into alternatives or into niches. Established cultural structures with deep power and control are likely to suppress subversive action. Such suppression has historically created an increasing wealth of alternative thinking over time. Open acceptance of plural values without critique leads to what might be called 'mass relativism', which if the idea were pursued might lead to high degrees of complexity or chaos. Probably the greatest threat to intellectual subversion and critique is political and cultural apathy. I am optimistic. Given the increase in reflexivity and global awareness of comparative cultures, such apathy is not apparent in the West, despite tendencies for 'dumbing down' of mediated ideas into plurally acceptable superficialities.

Futures Studies as 'Worldviews'

The world is a small place and wholly interconnected. Views of the whole world (systemic worldviews) are therefore axiomatic for thinking about the future of the world or about any holistic part of the world. Many views of the systemic worldview exist. Worldviews of the whole world, that is, views that embrace epistemological diversity, are difficult to attain. Two brief descriptions of 'events' I have recently experienced (and interpreted) illustrate this.

We are attending the UK Futures Group. We listen to Ziauddin Sardar, and to others, proclaim 'alternative' thinking, non-Western thinking. The listeners discuss these presentations for over an hour. Their discussion answers an unasked question: 'If you wanted to articulate a worldview (in futures studies), what elements would it cover and how would you do it?'

From the discussion, the list would include:

- Economic development dynamics;
- Futures thinking;
- Art (once we have broken through barriers of authorised expression);
- Economics (e.g. the concept of interest charges);
- Political representation and power;
- Social values in science and technology;
- Meanings of time; and
- Purpose and goals of life.

The list, which is evidently grounded in Western thought, illustrates how a 'worldview' is an exercise in reconceptualising others' experiences through your own ontological 'looking-glasses', language and metaphor.

The second example is a reflection on the global event of the World Futures Studies Federation (WFSF) in Brisbane, Australia, October 1997.

> It is nearly the end. There has been a lot of activity and sound-bites from many people and conversations on what we can do for future generations. Many cultures come together. There is an uneasy accommodation of hard-nosed planning, commercial exploitation of the futures genre, utopian visions, unselfish commitments and colonisation of the meek. A money collection from participants ensues to finance computers for WFSF members from African villages. You cannot colonise without computers and Internet connections.
>
> A teacher says of the plenary sessions of group feedback, 'well it was very interesting—there were some good bullet points . . . and there were some leaves'.

Pluralism, like most things, starts with awareness. We have a long way to go to assimilate plural cultural mores into futures studies. People at the WFSF were trying, but no one knows how, except by doing it. *In this domain, futures studies becomes the question 'In what ways can the problem be seen?'* First, the paths of the present need to be problematised, that is, turned into debatable assumptions rather than accepted truths. This problematising is common in futures studies, as in any questioning stance of any discipline. But taking a worldview demands infinite relativism in the formulation of the problematic and the abstracting of its vital features. This is impossible; epistemological worldviews of a worldview are impossible.

Futures Studies as Guardians of Future Generations

If scientific progress has given a meaning to futures studies, an equally powerful motive has also produced outstanding development in the field. The construction of a 'case' for alternative futures has been motivated by deep concern for the future effects of current trajectories. This motive should remain an important source of new ideas, and it takes meaning from, as well as giving meaning to, futures studies.

The mission of many 'alternative' futures groups, such as the anti-road lobby, is to act as guardians of future generations. Concern for the impacts of today's actions on future generations is a duty. It is a different concept from looking ahead at what might be. *In this domain, futures studies becomes the question 'What are the consequences of today's actions?'* To say that the future is foreclosed by the cultural colonisation of the West, or to say that futures studies as a discipline is unidimensional are examples of futures studies as a warning. Such discourse illustrates that thinking about the future impact of present actions is alive and well, although its visibility is limited. The idea of future guardianship as a duty of all people is not a mainstream value in Western society.

Such is the diversity of actions and contexts that 'create' the future that a separate discipline, such as futures studies, cannot take on the role of guardian of the future. Those engaged in substantive enterprises—all of us doing whatever we do—are creating exacting demands on future generations. Each of us needs to be enabled to use the skill of foresight in our actions. Cultural values underpin the judgements of future dangers. Guardianship cannot be left to marginal groups, but needs to be more central culturally, that is, something that people do. Futures studies has a role to play, for example to develop concepts of due diligence for future generations. A challenge for futurists, and for futures studies, is to understand and proclaim cultural ambition.

Underlying the notion that we are able to guard the future if only we are aware of the need is the idea that we can determine the future. Paradoxically, this notion does not withstand critique from the philosophical perspectives that identify the existence of subjective, interpreted meanings of 'objective' evidence. The notion of guarding the future is equally as difficult as that of scientific determinism of the future. How do we know that measures to guard against foreclosure or against a particular 'unwanted' path are valid? The same critiques of established thinking need to be applied to alternative thinking. Is this the role of futures studies? Should not foresight encompass modes of critique that continually press into service philosophical developments? Is *foresight*, and futures studies, a dissemination-action vehicle for comparative philosophy?

Futures Studies as Forecasting

The relationship between futures studies and forecasting is well documented. The issue here is the future of forecasting. As this is a massive area, I will focus on one main issue for the 'future' of forecasting. This is the emerging field of complexity and non-linear modelling. The relationships between complexity theories and futures studies have been recognised, in particular during the 1990s. The notion that causality is non-linear—that is, that co-evolution or interactions between interdependent parts create unforeseen and far-from-equilibrium outcomes—is not news to futurists. The whole notion of alternative outcomes as a principle guides the enterprise of futures studies. Similarly, the principle widely espoused by the futures studies community, that the future 'cannot be predicted', effectively forms a relationship between futures studies and forecasting by giving meaning to the products of forecasting processes.

The reason that complexity is relevant to the futures of futures studies is that our experience of organic, 'living' phenomena enables us to model their behaviour as systems. Social systems and ecological systems at many organic levels, from simple cells to rainforests, appear to us to behave in similar patterns. These patterns are complex, but they can be abstracted and modelled to produce simulations that behave 'as if' they were the abstractions from our observations. In these simulations, paths of behaviour and states of being change over time-phases, change through 'generations' of existence. The patterns of behaviour are not regularly predictable. For example, they do not follow a determined path to equilibrium nor repeat themselves exactly each time.

What complexity theory offers are analogies between different dynamical forms, providing models and metaphors to help in the understanding and interpretation of the 'real' characteristics yielded from research of specific phenomena. Analogies have limitations. For example, the evolution of a non-reflexive species is not the same as the evolution of a reflexive (human) species[5] and social systems have a different reality from natural systems.[6] The enterprise of complexity theory is to attempt to simplify rather than complicate. Essentially, complexity theory is a study of order, not of chaos. Futures studies is also a study of order, of alternative 'orders' that may exist, or may be understood to exist, beyond the present time.

What complexity theory mainly contributes to 'the future' are not direct analogies, metaphors or tools for alternative thinking. Its main contribution is to mainstream scientific and social science paradigms that illuminate the unpredictability of events and existence beyond the present.[7] It supports the 'alternatives' thesis of futures studies and opens a future debate on the guardianship of the future. *In this domain, futures studies becomes the question 'What is the pattern of regularities and what is the limit of the context of those regularities?'*

There is another aspect of modelling and forecasting that has a future. Forecasting and systems modelling, such as artificial life, intelligent materials and sensors are part of our existence. They do not stand outside our existence. They contribute to the 'self-fulfilment' of a forecast. The reach of 'machines' that incorporate present expectations or intentions for the future as instructions for the future is increasing. This is occurring outwardly into society through information and communication technologies (the Internet is a prime example) and inwardly into the human and other organic life through nanotechnology and genetics. The quotation below is from Christian Huebler, whose work *Knowbotic Research (KR + cF)* develops hybrid models of knowledge generation. These models are 'complex dynamic fields which produce an exchange between virtual agents, poetic machines and interactive visitors. They enable an observer to participate in the physical'. *In this domain, futures studies becomes the question 'How open to respond to novelty are the systems we depend upon?'*

> Our bigger concern is the topic of *Wirklichkeitskonzept.* With the term 'virtual reality' you can define a dimension that belongs to the computer. This is just play, but I think we play in accordance with rules of games dealing with phenomena that really have an effect on our personal life. The question is no longer what nature is, rather, what kind of nature do we want. We are embodied in the process of how nature merges, with the ability to go into the system and to change and manipulate it. We have to include in our research the term 'real' (das Reale), what comes out from the reality conception. We don't know if we really can discuss about Das Reale. It is a very delicate thing; in our work dealing with nature we must also deal with the economy and politics.[8]

Futures Studies as Planning

The word 'planning' is used here in the context of 'management', in particular of public and corporate investment. This meaning of planning is of intention and implies high degrees of predictability and therefore relatively short time-horizons. Futures studies as a discipline is sometimes located in university departments whose purpose is explicitly 'planning'; for example Departments of the Built Environment.

The impact of public planning and corporate planning spreads widely, and, as Tapio points out, the practice of planning is still dominated by 'discredited' linear systems thinking.[9] Of course, decisions have to be made. It is easy to confuse planning with futures thinking. The planning horizon for a decision is that time in which a decision can be implemented—no more. So, a decision to build an aircraft or supertanker needs a five-year horizon, because that is what it takes to build one. In Western economic-dominated cultures, the time-envelope for the decision also needs to account for the 'return' on the investment. This might be another ten or more years. Pubic infrastructure programmes may take a thirty-year time-horizon. These 'investment' decisions do not contemplate the whole-life impact of a decision. Does the oil industry's consideration of the long-run environmental costs of carbon fuels make any difference to their commercial decisions? The time-envelope of decision-making for most sponsors of planning-oriented futures thinking is not as long as the impact of the decision they make.

The futures studies for 'investment decisions' is about calculated probabilities, where control and risk-spreading are the issues, not the future beyond the time-envelope. The growth in 'risk management' as a discipline is testament to this focus, and to the growing imperative to 'manage' the unpredictability of future events. The 'management' response to unpredictability is to shorten response times ('time to market') and therefore to increase aggregate complexity and uncertainty.

Of particular interest to some of the Foresight Research Centre's stakeholders is the vexed issue of 'business plans'. These are the devices through which people running businesses, small and large, are expected to communicate to stakeholders their expectations of the future and their mobilisation of resources in relation to these expectations. Paradoxically, futures studies can give real value to these, though not often recognised. *In this domain, futures studies becomes the question 'What are the "waves" on which the enterprise's expectations are founded and what inter-relationship do the intended actions have with these?'* A 'business plan' can become a nexus of the articulation of the path of a relationship between the enterprise, its stakeholders and their mutual future relationship.

The most important relationship between planning and the future is that people, and Earth, will experience the outcomes of planning decisions. The key issue with respect to the futures of futures studies, in the context of inclusive views of the future, is the relationship not with modes of planning but with the legitimising of decisions that affect people not making the decisions. *In this domain, futures studies becomes the question 'If the futures of communities are navigable, then who decides the path and by what right?'* The issue of public or community consent, and of new forms of 'democracy', is vital; it has life in various political movements, for example in the communitarian movement influenced by Etzioni.[10] The realm of governance and power is a nexus for uncertainties, beliefs and actions about the future. If futures studies does not illuminate this, its 'headlights', as de Jouvenel has it, are pointing the wrong way.

Futures Business Futures

Futures is mainly an entertainment business. Few take it seriously. The images presented are mainly either reminiscent of 1950s science fiction or mystical. Why are stories about the future called 'science *fiction*', whereas predictions are framed as astrology and mysticism? Is the meaning of each of these actually reversed in Western culture? Most do not believe in the astrologer's predictions. We make them into what we want to believe. We can believe in the predictions of the science fiction writer, but feel no relationship between now and 'then'. Both sources of futures thinking bring us what we desire more than anything else, predictability. And if not predictability, then control. We want to believe. This is a powerful motive and makes futures into a business.

Many, if not most, people in Western cultures are comfortable with the belief that scientific discovery can solve the problems that science has created, as well as continuing to meet people's increasing expectations of material goods. Vast amounts of resources are utilised in scientific discovery, mainly without serious public governance. Most people do not understand science, and only recently has its public standing been under some pressure. We do not want to understand; we just want to feel secure. It is the same with the business of forecasting.

The clients of forecasting consultants want comfortable certainties or, at worst, comfortable uncertainties. My observations of major programmes of forecasting are that they are linear; that is, they say in effect 'more of the same'. For example, the UK Technology Foresight programme of 1995 declared what would be the key technologies of the near future so that they should have public R&D investment. This is not only linear but also self-fulfilling and self-serving to its active participants. These 'predictive' methods are planning, not foresight. How refreshing it would be to see an effective national foresight programme as an ongoing process of critical and creative endeavour, rather than something that produces 'consolidated findings'.[11]

So is being entertainment and a comforter the future for the business of futures? Below is a description of an event that occurred recently and that is part of the way we do our work at the Foresight Research Centre. To me it says much about the history of futures studies and something about its future.

There are about 30 of us gathered around a large table. We have spent the afternoon thinking about paths to the future and how these gather momentum and are signified through events. Now we face the fully featured presentation by a Western futurist, someone who has done futures in a corporate and establishment world for most of his distinguished career.

It is long. It is full of evidence. It is creative in its formulation of meaning. It asserts a future foreclosed by technological solutions and social and economic progress. It is an 'establishment' perspective on the future, promulgated to policy-makers in the UK and many other nation-states. It is, according to its promulgator, barely acceptable to many business people and policy-makers because of its bleaker aspects. The decision-driven community does not always assimilate these bleaker messages. In truth, the messages say 'more of the same'.

We slept on it, some fitfully after a too generous meal. This group of thirty people, mostly white, mostly professionals or entrepreneurs, mostly the great British Middle Class. This group rejected the linear technocratic future presented. This group rejected the paths of 'progress' so eloquently delivered. Many had admiration for the coherence of the analysis and prognosis. Most were impressed by the confidence that comes with years of finely honing ideas. It was not that the group rejected a particular proposition with respect to a definite future. I do not know quite what each person was rejecting. As a group there was a rejection of the certainty of the possible futures presented, of the dominant paradigm used in the analysis. The paradoxical rejection of something concrete and well formed. Paradoxical because the essence of Western futures as presented was implicitly that they must be certain, even if only by degree or probability.

The scene ends with healthy frustration. It is the culture of Westerners to have answers. The quest for knowledge is insatiable. Reflexivity of knowing and doing becomes doing itself in that quest. The frustration is that the more answers we have the less we know about what the answers mean. Liberal pluralism leads to such frustration. Pluralism increases uncertainty. The requirement for certainty by 'decision-makers' submerges pluralism. We know that this is not the way it needs to be. We are thankful to the presenter for enabling us to understand this.

So, there is your hope for the open, non-foreclosed, future. If a bunch of eclectic, mainly Western middle-class professional people reject linear technologically foreclosed futures, then does that make a difference? I think so. Not that they will make the difference, just that there is a difference.

We are at a turn of our understanding of progress and certainty. I have little doubt that in the current trajectory objectivism is the god. The consumer is important; consumption is important; the market facilitates exchange of objectively measured values. This simple value pair—the market and consumption—coupled with the communication society leads inexorably in the trajectory of greater consumption and hence greater power to those who its fulfilment. This trajectory has not reached the top of the S-curve (an invention of our linear futurists). *In this domain, futures studies become the question 'What, then, are signs or prescience of turns in this trajectory?'*

The Futures of Foresight

The Foresight Research Centre at Durham University Business School is developing as an 'Institute of Foresight', contributing to the social capacity for foresight suggested by Richard Slaughter.[12] We work mainly with small firms and the stakeholders of small firms.[13] Can we take a 'worldview' that encompasses the cultural diversity of this domain? We can only say what we see and think, and help others to do the same. What we 'see' is guided by our epistemological 'lenses'. Whatever we do, it will only be a perspective, or a finite number of perspectives on the domain of our stakeholders and the relationships between the past, present and the future.

In service to stakeholders, the role of research and theory is to illuminate meaning, making things mean something they did not before, where the value of the meaning emerges through discourse and action. Therefore our role is to develop, discover and share meanings with stakeholders. In undertaking these processes we need to be vigilant to the myopia of our interpretations, to the fallibility of our ideas and 'theories'. It is the practice of European academic researchers to display the principles underlying interpretations. This is in a sense 'normal' for academic research, though it sometimes difficult to sell the idea of fallibility to a commercial client expecting The Answer.

And what is it that we are contributing to our stakeholders, the people making a living from small enterprises? At a substantive level, to use Jim Dator's metaphor, we are part of our stakeholders 'surfing' process. Understanding the waves they are 'surfing' on and how to stay afloat, identifying and channelling emerging waves and tsunami. We are also concerned with their awareness of the meaning of future generations, in business, social and personal domains and the relationship that each enterprise has with these. We are concerned with the potential of emergence and co-evolution. And we are concerned with the concepts and cognitive tools to utilise in thinking about our joint relationship with the future.

Does that say to our stakeholders that the future is foreclosed by Western ideals, by Western philosophies, by corporatism, by science and technology? Does our reflexivity amplify 'Western' futures? Does the evidence we have from observations and analysis of our stakeholders and their environment tell us what, if anything, is foreclosing the future? My immediate reaction to this question is that our understanding develops as other people respond to our ideas; whether they critique, validate or reject, amplify, modify or otherwise act upon these ideas. Those whom we learn from shape the way we think and so shape our relationship with the future.

At the start of this chapter I implied that the more futures studies becomes a discipline and specialisation the less well it does its work in the development of foresight. The way towards a more open, global discourse on futures studies, in any and every domain, is to engineer a more open, global discourse on world futures. In this sense, the WFSF is a good example, but it is not enough. There is room for more, much more and in many different domains. This will change the future of futures studies. Who will do this? In my view it is part of the duty that comes with being a 'World Futurist', so includes those of us who profess our part of world futures. If there are non-Western world futurists, let them speak now.

Notes

1. Richard Slaughter (ed.), *The Knowledge Base of Futures Studies*, vol. 1, Hawthorn: DDM Media Group, 1996, p. 372.
2. E. Fuller, *Small Business Trends 1994–8*, 2nd edn., Durham: Durham University Business School, 1994, p. 144.
3. Demos Home Page, 1998 <http://www.demos.co.uk/index.htm>
4. Richard Slaughter, *The Foresight Principle*, London: Adamantine Press, 1995, p. 232.
5. P. Jeffery, 'Evolutionary Analogies', *Futures*, 28(2): 173, March 1996.
6. Arturo Escobar, 'Elements for a Post-Structural Political Ecology', *Futures*, 28(4), May 1996.
7. Gulbenkien Commission, *Open the Social Sciences*, Stanford, CA: Stanford University Press, 1996.
8. P. Atzori, 'Discovery CyberAntartic: A Conversation with Knowbotics Research', *Theory, Technology and Culture*, 19(1–2): article 38, 1996.
9. P. B. Tapio, 'From Technology to Participation? Positivist, Realist and Pragmatic Paradigms Applied to Traffic and Environmental Policy Futures Research in Finland', *Futures*, 28(5): 453–70, June 1996.
10. Amitai Etzioni (ed.), *Communitarian Thinking*, Charlottesville, VA: University Press of Virginia, 1995.
11. Office of Science and Technology, *Consultation of the Next Round of the Foresight Programme*, London: OST, 1998, p. 13.
12. Richard Slaughter and Martha Garrett, 'Towards an Agenda for Institutions of Foresight', *Futures*, 27(1): 91–5, January/February 1995.
13. Foresight, *Visions of the Small Business Landscape in 2006*, Durham: Durham University Business School, 1997. <http://www.durham.ac.uk/foresigh>

11
Futures Studies and Global Futures

JAN NEDERVEEN PIETERSE

These reflections seek to assess the importance to and ramifications for future perspectives of globalisation and to situate the theme of global futures in future studies. In the form of vignettes, these reflections address three types of question. First, why should global futures be addressed? Second, in what fashion can they be addressed, according to what kind of premisses? Third, what form could global futures take? It is argued that on account of globalisation and its ramifications, futures must take on global scope. On the one hand, there are no sustainable futures without some form of political regulation of economic, technological and ecological change. On the other, the only effective regulation is global regulation. North–South differences are of central importance to this global horizon.

Why Global Futures?

Why would global futures figure on the agenda? The answer to this question centres on the ramifications of globalisation, technological change and the gradual emergence of global citizenship. Globalisation refers to the accelerated worldwide intermeshing of economies, and to cross-border traffic and communication becoming ever denser. Technological change is speeding up. Risks and opportunities are globalising. All this belongs to everyday experience. Accordingly, globalisation means global effect, global awareness, and therefore increasingly it also means global engagement.

Citizenship under these circumstances is no longer simply national, or, more precisely, the national domain is now one among several relevant organisational spheres, and citizenship is becoming increasingly national *and* local, regional, global at the same time. In addition citizenship is no longer as state-centred as it used to be. The point of democracy is no longer simply influencing the actions of government. National governance has become one institutional sphere among several. Citizenship is not simply 'international' either, because the concerns at issue are not merely a multiplication of nation-state or intergovernmental structures. Governance is increasingly a matter of international politics, supranational institutions, international treaties and law, in the process involving macro-regional bodies, transnational corporations, transnational citizen groups, and media—interacting in complex, turbulent, multicentric ways.

Considered in an evolutionary context, humanity has been growing in capacity, technological accomplishment, and reflexivity. Collective awareness of concerns that affect the species and the planet—such as the environment, population, development—has been growing and so has their public articulation, notably in UN global conferences, so that arguably a global public sphere is emerging. At the same time, technological and political accomplishment and awareness do not 'line up' to add up to a condition of collective capacity. Interests are widely dispersed, subjectivities and agendas quite diverse, and institutional capacities relatively feeble.

Why futures?

Anticipation and planning used to be a prerogative and defining feature of government—'gouverner ç'est prévoir'. It extended to business and finance in tandem with the development of instruments of credit (banking, securities, options, derivatives) and insurance, which hinge fundamentally on the capacity to estimate, calculate and hedge outcomes. In both spheres, governance and business, forecasting has achieved considerable technical sophistication.

Recently the World Business Council for Sustainable Development initiated a project on Global Scenarios with the following justification:

> Planning for a sustainable future requires business to be able to anticipate and not just react to change. This is the rationale underlying our project on Global Scenarios. This project is designed to help businesspeople reach a shared view of the future and challenge the 'mental maps' they hold about sustainable development. This will allow them to anticipate, not react to, the exposures facing their corporations and ensure that they are fostering sustainable development.

Here the capacity to anticipate is presented as enabling for businesspeople: it puts them ahead in relation to circumstances, technologies and presumably also public criticism, and would enable them to develop a shared agenda.

Capacities to anticipate and plan are crucial to business and financial markets, to governments and international institutions. Accordingly, some futures have already been planned and negotiated, bought and sold several times over before citizens have even begun to think about them. This means that the horizons and agendas according to which futures are planned and designed reflect limited interests and agendas. Should such capacities to anticipate be reserved to business and government, or should they too be a matter of broad public awareness? If they did not become part of civic reflection it would mean that citizen groups, NGOs, would ever be relegated to a back seat—forever *reacting* to the futures designed, prepared and communicated piecemeal by governments and corporations.

At the same time, both government and business planning are constrained. Governments increasingly deal with many political and social forces and variables.

James Rosenau referred to this as turbulence and 'post-international politics'.[1] Corporations are exposed to such flux in the market-place that they operate with limited time-horizons. In the market-place contingency is a fact of life—thus, 'there is absolutely no way, in the evolving marketplace, that you can know exactly who the suppliers, customers, competitors and collaborators are'.[2] A standard quip in business management is: look at how many companies of the Fortune 500 still figure on the list five or ten years hence. It also follows that in government and business planning the command-and-control model no longer applies.

Democratisation of futures

The 'colonisation of the life-world'—commodification, bureaucratisation—is a familiar metaphor (although with the reconfiguration of the state, bureaucratisation is to an extent being replaced by privatisation and informalisation). A function of this is a routinised process of the colonisation of futures, as a consequence of the concentration of forecasting and planning capabilities in government and corporate hands. The ordering of public space, for instance urban planning and infrastructure planning, is an example. What is needed, then, is a decolonisation of futures or, to use more general language, a democratisation of futures. Gradually it is becoming a common understanding that not only the end stage of public planning but also the design stage needs to be participatory.

Citizen groups

There are many forums in which governments, international institutions, banks, corporations compare notes, set agendas and build coalitions. This happens less among citizen groups, except as 'alternative forums' taking place on the periphery of intergovernmental conferences or regional meetings (of the EU, APEC, NAFTA). What do take place are regional meetings of social organisations, sectoral conferences and academic conferences. Both anticipation and joint agenda setting are less developed among citizen groups than among corporations and governments. They are not as well endowed with think-tanks, nor do they organise forums aligning their views and agendas. Alignment is happening even less *across* areas of concern—for example, human rights groups comparing notes with environmental groups, environmental organisations comparing notes with women's groups and indigenous peoples, etc.—except locally. Citizen groups concerned with human rights, ecology, women or community issues all have their values and preferred futures. But where do they intersect, interconnect? How and to what extent do these various single-issue concerns and futures line up? Thus, for citizen groups a preoccupation with futures, local and global, would involve several functions: developing anticipatory sensibilities and capabilities; developing a proactive stance; dialogue across groups; aligning normative concerns; making futures a matter of public concern.

Ideological futures

Another reason why futures thinking is taking on a new relevance is because several modes of anticipation that were available in the past have lost their relevance or appeal. Futures used to be packaged and delivered as part of the grand ideologies that framed the social and political movements of the past, particularly nationalism and socialism. The 'national question' and the 'social question' of the 19th century reconverged in different combinations in the 20th-century social movements, such as the anti-colonial national liberation movements.

The ideologies bequeathed by the 19th century followed positivist epistemologies and involved structuralist modes of thinking and analysis—relating to macro-processes such as imperialism, capitalism, dependency. The future scenarios that emerged from these are now no longer viable or attractive. Nationalism is making place for postnationalism, or at least for a trend of reconfiguration of energies in various directions—local, regional, macro-regional, international, global. Delinking or dissociation from capitalism has little meaning in a real world where localities scramble to attract foreign investment. The expectation of a world-scale crisis of capitalism, followed by an opening towards socialism, has now very few adherents (among the last are world-system theorists).[3]

Regional futures

Samuel Huntington's 'clash of civilisations' seems so hopelessly static and antiquated that even arguing against it feels like a waste of time. If civilisational destinies or regional projects still seemed relevant a few decades ago, they now sound increasingly quaint. Calls for 'the West' to be concerned with this or that (as with Huntington) are outdated at a time when, among others, business is eyeing 'emerging markets' across the seas. Interpretations of India centred on 'Indic civilisation' have historical purchase and thus inform futures; but would they be sufficient to generate relevant future scenarios, or would they rather feed chauvinist Hindutva ideology?

Evocations of an 'Asian Century' sound outdated already before it has begun. First, the dynamics are Pacific rather than Asian; witness the intercontinental sprawl of the Chinese diaspora. Second, while there is an 'Asian Renaissance',[4] its lineages are not purely Asian (just as the makings of the European Renaissance extended well beyond Europe). Third, much talk of 'Asian values' is authoritarian in intent and moreover often a laughing-stock inside the region. Fourth, Asian industries, those of Japan included, depend on technology from outside the region, particularly the United States. Fifth, because of their export orientation East and Southeast Asian economies depend on markets in the West. Complementarities among Asian economies, while considerable and growing, would not be sufficient to sustain the countries' exports. Sixth, intercultural exchange—between East and West—is so far advanced and so deeply historically layered that in many ways the two can no longer be meaningfully separated. Seventh, in a 1996 speech in Beijing Prime Minister Mahathir of Malaysia declared the 21st century a 'Global Century' rather than an 'Asian Century'. This was a sensible gesture of diplomacy—eventually Western markets would not react well to an upsurge of inward-looking Asian chauvinism; it was also an expression of an Asian humanism that genuinely sets forth a global engagement.

Similar considerations apply to Islamic projects. The wide world of Islam represents an alternative globalism with considerable historical and civilisational depth, geographical scope and growing economic opportunities. At the cusp of the millennium, however, the Islamic world is not independent in science and technology, in investments and growth opportunities, in armaments and security. Financially and economically, culturally and politically, it is profoundly wired to global centres.

In 2020 Islam will be the second major religion in most of Western Europe (in some countries it already is). Some parts of Europe are now reindustrialising thanks to Korean and Taiwanese investments. Japanese management techniques offer a model to overcome the Taylor model of standardised mass production, from the Pacific to Ireland.[5]

Regional and civilisational projects are best and most intelligently viewed not as contradictory to but as part of global dynamics. Accordingly, regional projects should both inform and be informed by global futures.[6] This is not an argument for *going global tout court*. Rather, it is to argue for an interdependence and balance of local, national, regional and global engagements.

How Global Futures?

Utopianism

Utopias in the sense of blueprints of desirable societies no longer match contemporary mentalities. Critique of utopianism itself is now part of contemporary reflexivity. The utopias of the 19th century tend to reflect a similar positivist epistemology, authoritarian design and command-and-control outlook to the ideologies of the same epoch, which they elaborate. Apocalyptic mentalities find adherents particularly towards the turn of the millennium; witness the Davidians in Waco, Texas and the Aum Shinrikyo cult.[7]

Middle-ground futures

Having grown up during the Cold War, we have grown accustomed to polarised worldviews. Does not everything have a colour and does not every colour figure somewhere on the spectrum? If now, however, we consider the everyday practices lived and advocated by diverse circles, they are almost without exception based on synergies among state, market and social forces. The terms and forms of co-operation vary, but the old antagonisms and antinomies—between capital and labour, state and society—no longer survive as such. The middle ground, long shunned as tainted and suspect, is coming into its own. For instance, in the words of George Soros, 'Instead of there being a dichotomy between open and closed, I see the open society as occupying a middle ground, where the rights of the individual are safeguarded but where there are some shared values that hold society together'.[8]

Futures

Over time discourse has shifted from 'the future' to 'futures'. Forecasting implies a future that already exists 'out there'. The main problem is its visibility, which may be approximated by means of extrapolation and trend analysis. Presently the understanding is more in futures as options and opportunities. The main pre-occupation in future-thinking—certainly in business management—is no longer forecasting but *imagining* and secondly *creating* futures. 'The big challenge in creating the future is not predicting the future. It's not as if there is only one future out there that is going to happen, and that the only challenge is trying to predict which of the potential futures will actually be the right one. Instead, the goal is to try to imagine a future that is plausible—the future that you can create.'[9] This implies that futures are *open*—'there is no proprietary data about the future', 'nobody owns the twenty-first century'.[10] A corollary to this understanding of multiple futures is the now common technique of devising *scenarios*, in the sense of schematic representations of clusters of future options.[11]

Fallible futures

The openness of futures implies that they are premissed on human fallibility. Following Popper, this is what George Soros regards as basic to an open society. This also means open in terms of their view of human nature. Gillian Slovo concludes a book about her parents, Ruth First and Joe Slovo, on this note: 'I'd realised that memory, experience, interpretation could never be fixed or frozen into one, unchanging truth. They kept on moving, relentlessly metamorphosing into something other so that the jagged edges of each fragment would never, ever slot together.'[12] This articulates a sensibility that is much closer to our contemporary everyday sense of social experience than the deterministic and reductionist thinking of the past. 'Linearity is an artificial way of viewing the world. Real life isn't a series of interconnected events occurring one after another like beads strung on a necklace. Life is actually a series of encounters in which one event may change those that follow in a wholly unpredictable, even devastating way.'[13]

George Soros accepts part of the logic of laissez-faire thinking—'if our understandings are imperfect, regulations are bound to be defective'—but he disputes the conclusion that 'since regulations are faulty, unregulated markets are perfect'.[14]

Flawed regulation that is reflexive in relation to its flaws, then, is better than no regulation at all. We might term this *postmodern regulation*. It refers to an approach to regulation that is flexible in understanding the necessary though limited status of regulation in relation to social life, which is multidimensional, messy and reflexive. Regulation alters the field it seeks to regulate, generates loopholes, avoidance behaviour, resistance. Full transparency is an illusion—it was in relation to societies (as communist bloc countries showed) and it would be in relation to global conditions. Accordingly, global regulation must be bold in acknowledging the importance of setting global standards, and modest in recognising that the importance of regulation lies in part in the avoidance it creates. Thus, regulation should follow fuzzy logic rather than linear thinking.

What do futures feel like?

Futures are many. Every epoch has its futures. Every place has its futures. And within each place and period, different groups perceive different futures as they experience different hopes and fears. Futures are not only rational projects but also emotional experiences. Futures are not simply a matter of rational choice; they are made up of images, aspirations and anxieties, some of which are unconscious, escape or resist rationalisation. To futures there are both explicit and implicit dimensions, above and below the waterline, and not all that is implicit can be made explicit. Logic and plausibility play a part in choosing futures but so do emotional, aesthetic and imponderable considerations. So a relevant question to ask is, What do futures feel like?

Recent futures past

Generalisations are superficial, but at least we can review the dominant, hegemonic futures during the recent past. Thus, in the 1950s futures felt *modern*, like modern furniture and Le Corbusier architecture. A mainstream future was the American Dream, broadcast by means of Hollywood movies. By way of counterpoint there were dissident futures such as George Orwell's 1984 and Aldous Huxley's Brave New World. Futures thinking at the time—the birth time of 'futurology'—was centred on technology, such as space technology, armaments, robotics, artificial intelligence, and modernist in *esprit*. Some of these trends had been in evidence well before that—as in Jules Verne and science fiction, Italian futurism, Soviet Taylorism, and Gramsci's 'Americanism' and Fordism.

The backdrop of the Cold War provided a duopoly of futures—the Free World of mass consumption and the worker state of socialism. The rift between them intensified with the Vietnam war and '1968'. Mainstream futures were shaken and a wide array of dissident futures emerged, images of struggle and liberation—such as '*tier mondisme*', the civil rights movement, 'protracted people's war', the Great Cultural Revolution, Che Guevara's focismo, liberation theology.

Major concerns during the Cold War were peace and security. Security issues and strategic thinking played a large part in mainstream future-thinking (for instance, domino theory) and in dissident futures on the part of the peace movement (for instance, the idea of 'deadly connections' between Third World conflicts and superpower rivalry). The nuclear age and collective security cheerfully premissed on 'Mutually Assured Destruction' provided an aura of threat and doom, an aura of 'exterminism'. In the course of the 1980s this has been taking different turns, and 1989 sealed the demise of state socialism. In the words of Jim Clifford, 'I grew up in the everyday fear of this implosion and the real possibility that I and everyone I knew might not survive. The fear, a fact of life for more than three decades, has receded. I, my family, and my friends will probably live into the next century—a time with its own dangers, known and unknown, but at least without the threat of imminent extermination. All at once, the millennium feels like a beginning.'[15]

At the same time, from the 1970s a new range of futures emerged. Earlier studies of population growth had offered doomsday scenarios, neo-Malthusian premonitions of limits to population growth. Now environmental scenarios predicted *Limits to Growth.* The Club of Rome's report came out in 1972. It was followed by a host of studies, such as *Mankind at the Turning Point,*[16] *Building a Sustainable Society*[17] and the US-government-sponsored study on 'Global 2000'.[18] Also economists turned to futures thinking, such as Herman Kahn and the Hudson Institute,[19] the UN-sponsored project of Wassily Leontieff and associates[20] and the Interfutures project of the OECD.[21] Sociological futures typically addressed the relationship between technological change and socioeconomic, political and cultural transformation, in particular how the 'third industrial revolution' of information technology is changing social practices. Examples are Daniel Bell's postindustrial society, Alvin Toffler's *Future Shock* and *Third Wave,* Naisbitt's *Megatrends,* and Kenneth and Elise Boulding's work.

Gradually these futures—based on technology, developed by demographers, security analysts, environmental studies, economists, sociologists—have become part of general futures thinking. In the 1990s different tensions and futures have come to the fore. Identity politics and new social movements have taken the place of national and social struggles. All these have come into a new focus in opposition to free market politics. Globalisation became the new arena. The new terrain of contestation became neoliberalism, with structural adjustment in the South and the erosion of welfare states in the North.

Which Global Futures?

Neoliberal futures

As an ideology neoliberalism is probably past its peak. The trust in the 'magic of the market-place' that characterised the era of Reagan and Thatcher has run its course. Yet the criticisms of 'the market rules OK', while common and widespread, have so far not crystallised into a cohesive alternative perspective—except for 'pragmatism rules OK', which in effect means muddling through. Alternatives to neoliberalism reflect a variety of interests and positions that have not been able to cohere either ideologically or institutionally. Under the circumstances, monetarism is the default ideology and policy. Institutionally, in bodies such as the WTO and the IMF, neoliberalism remains the conventional wisdom. In development politics it prevails through the 'Washington consensus'. In NAFTA, it prevails in principle. In European politics, it prevails through the European Monetary Union. Financial and monetarist regimentation is ironing out the actual varieties of capitalism.[22]

This reflects the continuing hegemony of finance capital and the central position of financial institutions. The core of the neoliberal powerhouse is finance capital. Money and finance are the central arbiter and regulator of regional and global development. Ironically so, since, at the same time, finance itself is the most unregulated of all economic spheres—witness the growth of the hyperspace of international finance and its volatility. While monetarism serves as the default discipline and ideology of the neoliberal world, the financial world itself is out of control. The nexus between international trade and finance has been severed, so that speculation on currency fluctuations and other financial instruments has become quantitatively more important than production and commerce. Twenty-four-hour trading and electronic triggers have increased volatility and risk. Currencies are unstable; witness the Mexican peso crisis. The relationship between profit and taxation is eroding, or being reversed. Corporate tax rates are shrinking. Government subsidies and incentives take the place of taxes. Deregulation and access to hyperspace enable corporations to register headquarters and record profits at offshore tax havens.

One of the neoliberal futures of the 1990s is Kenichi Ohmae's *Borderless World*, or the world as a duty-free store.[23] This is a replay of the American Dream (which in the United States, if only because of work loads, is a dream no longer), a global extension of Walt Rostow's 'stage of high mass consumerism'. Kenichi Ohmae's former position as director of McKinsey Japan accounts for much of the signature of his perspective. This is a very 'Pacific' kind of ideology—American marketing and mass consumerism coming to Asia, and Japanese corporate strategies boomeranging westward.

The future of a borderless world for capital is gradually being turned into a self-fulfilling prophecy through the structural reform policies initiated by the IMF and the World Bank. It is being translated into a global dynamic by the WTO and the attempts to arrange for a future of free movement not only of commodities but capital as well. It is a world viewed from a Northern window, seen through Western and Japanese eyes. It fills in the blanks in Fukuyama's *End of History*, which predicts indefinite political stability in the advanced world while small wars and skirmishes sputter on in the periphery. Daniel Moynihan's *Pandemonium* of ethnic conflicts and Robert Kaplan's slide into anarchy in the periphery, from Africa to the Balkans, supplement this prognosis, whose regional effects are to be contained by means of selective 'humanitarian intervention'.[24]

Neoliberal futures are being contested on many grounds—the environment, labour, the right to development, local interests, cultural diversity. Looking back on the Rio conference on the global environment, Martin Khor sees a 'clash of paradigms': 'The free market paradigm . . . represented by the Bretton Woods institutions, which persisted in promoting structural adjustment programmes based on market liberalisation, and by the GATT/WTO which was dominated by the Northern governments advocating the opening up of markets (especially of developing countries)' and the 'paradigm of partnership and cooperation . . . represented by the United Nations series of world conferences'.[25]

This kind of characterisation, though not without plausibility, may be misleading. First, the concept of paradigm is derived from the natural sciences and either does not apply or would have a much looser meaning outside it. Second, it homogenises positions and suggests more coherence in positions than actually exists. It conceals the improvised and patchwork character of actual policy frameworks. Third, it sidelines the question of whether the 'paradigm' would be capable of reproducing itself, in other words, whether it has a future at all. Fourth, juxtaposing the two policy frameworks as paradigms gives a mistaken impression that they are somehow of equal status. Fifth, the image of a clash ignores the overlap in positions, eliminates the middle ground, and overlooks actual and possible cooperation.

The question is whether the neoliberal regime is capable of reproducing itself. Major elements of instability in the neoliberal scenario are: (1) financial instability—witness the growth of the hyperspace of international finance[26] and financial turbulence in Asia; (2) the risk of global oversupply;[27] (3) the problem of automation and 'jobless growth';[28] (4) environmental risk;[29] and (5) tensions between the market and democracy.[30] For such hazards, deregulation, liberalisation, privatisation provide no remedy.

Accordingly, some form of regulation is in the interest of parties, large and small. Without taxation, no infrastructure. Without taxation, no proper education, no affordable health care. Without taxation, no public sphere. Without a public sphere, no legitimacy, no security. This in itself is a familiar future scenario. It could take the form of a process of global 'Californianisation' while California is being 'Brazilianised'. A world of gated communities, high barbed-wire fences, steep hierarchies, robots at the workfloor and at the gate, and automated surveillance all round. A world of creeping privatisation of public space, as in Los Angeles (as in Mike Davis's *City of Quartz*).[31] A world of 'post-human development',[32] of which we see glimpses in *Blade Runner* and other cinematic dystopias.

This world, however, cannot reproduce itself. The neoliberal scenario, if all the padding of state support really is removed, is not coherent, is self-contradictory and self-defeating. Laissez-faire and the 'self-regulating market' have been *critical* positions (Manchester school criticism of mercantilism, supply-sider criticism of Keynesianism), but they do not add up to a stand-alone self-sustaining model.

Regulation in order to function must be of *global* scope, for anything short of global regulation invites evasion and 'dumping'. At the same time the development gap among countries at different levels of development makes the likelihood of any straightforward global consensus quite small—not for lack of trying; witness numerous sensible commission reports; witness the recurrent stalemates in global conferences on virtually any issue—the environment, population, trade. The real world stretches all the way from palaeolithic hunters and gatherers (Amerindians in the Amazon, Aborigines in the Australian outback, Khoisan in the Kalahari) to the high-tech worlds of information technology. In world music, cinema (John Boorman's 'Emerald Forest', Peter Weir's 'Last Wave', Jamie Uys's 'The Gods Must Be Crazy') and science fiction, in indigenous peoples networks and human rights forums, these worlds meet, but where else? The scope and limitations of global rendezvous have been demonstrated in the global conferences in Rio, Vienna, Copenhagen, Cairo, Beijing, Istanbul. These have been exemplary in the range of parties represented—international institutions, governments, TNCs, NGOs, unions, professional associations, media. But participation has hardly been on equal terms, and outcomes, on the surface at least, have been uneven and meagre.

Global development

A familiar friction runs between the 'Washington consensus', embracing the IMF and the World Bank, and the UN institutions and their agenda of human development (UNDP, Unicef, Unesco), social development (the Copenhagen Social Summit, UNRISD), sustainable development (UNEP). This rift—in short, between Washington and New York—goes back to the institutional divergence between the Bretton Woods institutions and the UN system that emerged after the second world war and took the form of different voting systems. One set of institutions would deal with the 'hard' issues of finance and economics, and the other with the 'soft' issues of social welfare, entitlements and human rights.[33]

Such a division of labour is now no longer tenable, because it is out of synch with current insights into how economics works—the hard and the soft are deeply interwoven. *Soft* elements such as education, health care, housing, income distribution, cultural exchange, social capital, civic trust and institutional density translate into *hard* economic data of productivity and growth. Cultural diversity can be viewed as an engine of economic growth.[34] Without community participation development projects don't work. Facilities such as micro credit—popularised by the Grameen Bank in Bangladesh and now also being applied in the United States and Britain—typically bridge the two spheres of concern. In relation to environmental concerns, the distinction between soft and hard does not make sense. In sustainable development—the lead paradigm of the 1990s—the hard and the soft cannot be separated. Thus, over the years the World Bank has been incorporating 'soft' elements as part of its brief—gender, participation, NGOs, environment; and also the IMF is engaging in dialogue with civic associations.

There is a broad spectrum of opinion in favour of recombining the Bretton Woods institutions and the UN system, with a view to realigning economic financial regimes and social development. According to ul Haq, with some display of diplomacy, there is no contradiction between structural adjustment and human development; it would just be a matter of designing the right policy mix.[35] But he also argues for recombining the institutions. One proposal is for an Economic Security Council that would take over the role of the International Financial Institutions under UN auspices;[36] another is for the adjustment of voting systems.[37] What matters, beyond the institutional arrangements, is the realignment of international financial regimes and social development. What is at issue is the role of central banks, particularly in the North, and their relationship with the IMF. The broader issue is the restoration of banking as a public utility, which ultimately involves a new mode of regulation.

Global governance

Seemingly slow progress with regard to global governance should not conceal the fact that we inhabit a world shaped by over a hundred years of international regulation and institution building. This includes arrangements with regard to time zones (Universal Standard Time), the International Red Cross, regulation of the conduct of war, UN treaties from the Declaration of Human Rights to the Law of the Sea, and the International Court of Justice. An international public sector already de facto exists. In this context sovereignty need no longer be thought of as a zero-sum game. Sovereignty is not reduced by sharing it. Another stepping-stone in the process of global governance is regional governance (as in the European Union). Trade-offs of pooling sovereignty include security and stability, reduced military spending, reduced conflict and anxiety, and economic and technological co-operation.

The Commission on Global Governance is at pains to point out that global *governance* is not global *government*.[38] It is not about creating a superstate but about strengthening the international order and international law. If we combine this with current ideas about politics and the state,[39] one can think of facilitative governance, or governance as management of networks and co-ordination of synergies across sectors.

Progressive agendas include 'double democratisation' within societies and in international relations, and 'cosmopolitan democracy', including the formation of regional parliaments,[40] the formation of world parties,[41] and substantive UN reform—'redefining the United Nations as an organization not of governments but, in the final analysis, of "the people" '.[42] What is at stake in global governance, besides its institutional design, is the question of global regulation.

Global regulation

Regulation in order to succeed must be global in scope, for anything short of that invites evasion. In order to achieve global scope regulation must be acceptable across North–South differences. It must encompass a politics of development that addresses the interests and agendas of advanced countries, NIEs and LDCs. In effect this means an agenda of 'global development'. Advocating global regulation on moral and ethical grounds, or purely political grounds, has no more appeal than the reach of ethical and political consensus. It is probably more sensible to treat global regulation primarily as a form of global risk management. Partnership develops when its advantages outweigh its costs. 'Mutual interest' then remains the guiding principle, but this should now encompass wider concerns, such as environmental and security hazards.

If straightforward negotiations do not deliver because the parties are unequal and have diverse interests, *widening* the terrain of negotiation to broader fields of common concern may open up new opportunities for give-and-take. Areas that come to mind are ecology, natural resources, regional security, migration, cultural diversity, indigenous knowledge. Exploring the rainforest for medicinal herbs while granting intellectual property rights to local inhabitants and contracts for revenue-sharing are an example.[43]

Not only must fields of negotiation be wider than they are at present, future proposals must also be interactive multidimensionally; that is, global governance must heed the global economy; global taxes must interact with policies regarding development, ecology, population, gender, cultural diversity. In other words, global futures thinking must turn on intersectoral synergies. This goes against the grain of the *sectoral* structure of institutions, ministries, and agendas. Sectoral barriers contribute to the stalemate in negotiations. In bureaucracies, as well as in the disciplinary structure of academia, they separate finance, development, welfare, ecology, and so forth, as part of the legacy of 18th- and 19th-century rationalism and its classificatory spirit. By definition they exclude from view and from discussion questions such as global governance and global taxes, which have profound multiplier effects across many spheres. These fences and boundaries are out of step with contemporary increasingly holistic and interdisciplinary understandings of development and politics.

At the same time, many conflicts of interest cannot be resolved within narrow frameworks. An example is the friction between the labour standards movement and the right to development. In the North labour standards (labour rights, no child labour, no sweatshops, minimum wage, union rights etc.) are widely viewed as a major terrain of progressive intervention. This position is shared by a broad coalition from trade unions to social organisations, from the ILO to the US Congress. In the South, however, this position is much more controversial. Many view it as undermining the competitiveness of the South, which is based on labour flexibility. For late-late industrialisers the competitive edge of low labour standards and minimal environmental regulation offers a slim chance to step into the industrialisation process.

Positions in the North are inconsistent in that while this attention is focused on labour, it goes together with measures and proposals to free the movement of capital. Freeing capital while regulating labour may not be a particularly even-handed way of going about things. Positions in the South are one-sided in that industrialisation on the basis of 'primitive Taylorisation' and environmental devastation does not add up to an attractive or sustainable development path. The problem is that two cases and levels of argument are intertwined. One is a general case for labour standards as a protection for labour at a time when the mobility of capital is increased; and the other is a particularist case on the part of trade unions and governments in the North facing capital flight and job loss, who seek to reduce capital flight by reducing the advantages of relocation. A level playing-field for labour is one question; regulating capital movement is another; comparative competitiveness is a third; the right to development is a fourth. Considering each in its own sphere, these conflicts of interest are insoluble. They could only be addressed within a wider, comprehensive approach.

Global social contracts

Often presented as a framework for addressing wide-ranging issues are new social contracts.[44] An advantage of the contract approach is that contracts are compatible with market standards as well as bureaucratic procedures; they cut across spheres of government, market and society. Contracts can be comprehensive and address multidimensional issues, while avoiding juggernaut and politically unrealistic options such as a 'new constitutionalism'. According to the Group of Lisbon, 'A contract is the appropriate choice when the parties involved are numerous, the problems are complex and multidimensional, and the solutions are of a long-term structural nature.'[45] The Group of Lisbon proposes global social contracts on basic needs, democracy, culture and the environment. To serve as settlements across sectors or domains, they could take the form of specific targeted negotiating platforms, such as *global taxes for global information structure, Tobin taxes for restraining international financial speculation, ecotaxes for planetary survival.*[46]

Implementation

In business, the talk is not simply about *imagining the future* or *strategy*, but also about *implementation* or *building the future.*[47] Governments and international institutions do not simply imagine futures either. Malaysia's 'Vision 2020' (catch up with advanced countries) serves as a government policy target. This kind of constructive attitude—minus the authoritarian attitudes—would be welcome also on the part of citizen groups.

Among the avenues out of the global stalemate are: (1) a stock-taking of socially progressive best practices and future proposals—this refers to innovation not simply in a technological sense, but in social practices, institutions, values and expectations; (2) articulating them, stitching them together across domains and geographies; (3) the development of global public opinion concerning planetary predicaments and future options; and (4) the development of multidimensional relations of negotiation, that is, across different sectors and dimensions.

The development of global public opinion is necessary for the generation and articulation of political will. Besides, if the objective of progressive futures is democratisation, the means towards their realisation must be democratic as well. First, the infrastructure is increasingly available. This is the upside of 'CNN culture' and planetary satellite wiring. Second, there have been breakthroughs in recent years, for instance with regard to the environment, human rights and women's rights. In a fairly short time-span the environment has become recognised as a planetary concern and institutionalised in the notion of sustainability and sustainable development as the yardstick of all initiatives, economic and technological. Human rights and women's rights are other fields in which considerable social progress has been achieved. No matter, then, the range of complexities, political sensitivities and cultural differences, social progress can be achieved. (Other breakthroughs have not been the consequence of global public opinion. An area of profound change in recent years is international security. Here the major source of change has been structural changes in international relations—the break-up of the USSR and the end of the Cold War—which have been partially unanticipated.)

These avenues may be viewed as obstacle courses of complexity. However, what we now find unimaginable may well be commonsense and common practice a few years or decades hence. In the course of this single century there have been so many changes that had been unthinkable not too long before. Various future options are interdependent. Global regulation and global taxes cannot be implemented without further progress in global governance. Global governance depends on global public opinion and political will. One way of seeing this is as a stalemate in which progress in any sphere cannot proceed without change in other domains. Another view is that this is a virtuous circle in which progress in relation to one sphere means progress in relation to all.

Notes

1. J. N. Rosenau, *Turbulence in World Politics*, Brighton: Harvester, 1990.
2. C. K. Prahalad, 'Strategies for Growth', in R. Gibson (ed.), *Rethinking the Future*, London: Nicholas Brealey, 1997, pp. 62–75 (p. 66).
3. e.g. I. Wallerstein, 'Peace, Stability, and Legitimacy 1990–2025/2050', in G. Lundestad (ed.), *The Fall of Great Powers*, New York: OUP, 1994; Samir Amin, *Capitalism in the Age of Globalization*, London: Zed, 1997.
4. Anwar Ibrahim, *The Asian Renaissance*, Singapore: Times Books, 1996.

5. P. Walley, *Ireland in the 21st Century*, Dublin: Mercier Press, 1995, pp. 150–1.
6. Cf. Charles Oman, *The Policy Challenges of Globalisation and Regionalisation*, Policy Brief No. 11, Paris: OECD Development Centre, 1996.
7. D. Thompson, *The End of Time: Faith and Fear in the Shadow of the Millennium*, London: Sinclair-Stevenson, 1996.
8. George Soros, 'The Capitalist Threat', *Atlantic Monthly*, February 1997: 9.
9. Gary Hamel, 'Reinventing the Basis for Competition', in Gibson, *Rethinking the Future*, pp. 76–92 (p. 81).
10. Ibid. 81; R. Gibson, 'Rethinking Business', in Gibson, *Rethinking the Future*, pp. 1–14 (p. 6).
11. An approach that was first implemented in the Interfutures project of the OECD, *Facing the Future*, Paris: OECD, 1979.
12. Gillian Slovo, *Every Secret Thing: My Family, My Country*, New York: Little, Brown, 1996.
13. Michael Crichton, quoted in Gibson, 'Rethinking Business', p. 6.
14. Soros, 'The Capitalist Threat', p. 6.
15. James Clifford, *Routes: Travel and Translation in the Late Twentieth Century*, Cambridge, MA: Harvard University Press, 1997, p. 344.
16. Milhaljo Mesarovic and Eduard C. Pestel, *Mankind at the Turning Point*, New York: E.P. Dutton, 1974.
17. Lester R. Brown, *Building a Sustainable Society*, New York: W.W. Norton, 1981.
18. Council on Environmental Quality, *The Global 2000 Report to the President*, Washington, DC: Government Printing Office, 1981. These forecasts are discussed in B. B. Hughes, *World Futures: A Critical Analysis of Alternatives*, Baltimore, MD: Johns Hopkins University Press, 1985.
19. Herman Kahn, William Brown and Leon Martel, *The Next 200 Years*, New York: William Morrow, 1976.
20. Wassily Leontieff *et al.*, *The Future of the World Economy*, New York: OUP, 1977.
21. OECD, *Facing the Future.*
22. M. Albert, *Capitalism against Capitalism*, London: Whurr, 1993.
23. K. Ohmae, *The End of the Nation State: The Rise of Regional Economies*, New York: Free Press, 1995; K. Ohmae, *The Borderless World: Power and Strategy in the Global Marketplace*, London: Collins, 1992.
24. D. P. Moynihan, *Pandemonium*, New York, Random House, 1992; Robert D. Kaplan, *The Ends of the Earth: A Journey at the Dawn of the 21st Century*, London: Macmillan, 1997. For a critical discussion of humanitarian intervention see J. Nederveen Pieterse (ed.), *World Orders in the Making: Humanitarian Intervention and Beyond*, London and New York: Macmillan and St Martin's, 1998.
25. M. Khor, 'Effects of Globalisation on Sustainable Development after UNCED', *Third World Resurgence*, 81/82: 511, May/June 1997 (p. 9).
26. H. Wachtel, *The Money Mandarins: The Making of a Supra-national Economic Order*, rev. edn., London: Pluto, 1990.
27. W. Greider, *One World, Ready or Not: The Manic Logic of Global Capitalism*, New York: Simon & Schuster, 1997.
28. J. Rifkin, *The End of Work*, New York: Tarcher/Putnam, 1995.
29. H. E. Daly and J. B. Cobb, Jr., *For the Common Good*, 2nd edn., Boston: Beacon Press, 1994.

30. Jacques Attali, 'The Crash of Western Civilization: The Limits of the Market and Democracy', *Foreign Policy*, 107: 54–64, 1997.
31. M. Davis, *City of Quartz*, London: Verso, 1990.
32. Mike Featherstone, 'Beyond the Postmodern: Technologies of Post-human Development and the Question of Citizenship', Global Futures public lecture, The Hague: Institute of Social Studies, 1997.
33. H. W. Singer, 'Rethinking Bretton Woods: From an Historical Perspective', in M. J. Griesgraber and B. G. Gunter (eds.), *Promoting Development: Effective Global Institutions for the 21st century*, London: Pluto, 1995, pp. 1–22.
34. K. Griffin, 'Culture, Human Development and Economic Growth', Working Paper in Economics 96-17, University of California, Riverside, 1996.
35. M. ul Haq, *Reflections on Human Development*, New York: Oxford University Press, 1995.
36. Cf. Report of the Commission on Global Governance, *Our Global Neighbourhood*, Oxford: OUP, 1995; H. Henderson, *Building a Win–Win World*, San Francisco, CA: Berrett-Koehler, 1996.
37. R. Green, 'Reflections on Attainable Trajectories: Reforming Global Economic Institutions', in Griesgraber and Gunter, *Promoting Development*, pp. 38–81.
38. Group of Lisbon, *Limits to Competition*, Cambridge, MA: MIT Press, 1995.
39. e.g. G. Mulgan, *Politics in an Antipolitical Age*, Cambridge: Polity, 1994.
40. D. Held, *Democracy and Global Order*, Cambridge: Polity, 1995.
41. W. P. Kreml, and C. W. Kegley, Jr, 'A Global Political Party: The Next Step', *Alternatives*, 21: 123–34, 1996.
42. Y. Sakamoto, 'Civil Society and Democratic World Order', in S. Gill and J. Mittelman (eds.), *Innovation and Transformation in International Studies*, New York: Cambridge University Press, 1997, p. 8.
43. L. Van der Vlist (ed.), *Voices of the Earth: Indigenous Peoples, New Partners and the Right to Self-determination in Practice*, Utrecht: International Books, 1994.
44. e.g. Report of the Independent Commission on Population and Quality of Life, *Caring for the Future*, New York: OUP, 1996; Rifkin, *The End of Work*.
45. Group of Lisbon, *Limits to Competition*, p. 110.
46. On global taxes see e.g. D. Felix, 'The Tobin Tax Proposal: Background, Issues and Prospects', in H. Cleveland, H. Henderson and I. Kaul (eds.), *The United Nations: Policy and Financing Alternatives*, New York: Apex Press, 1995, pp. 195–209. On 'infostructure' see M. Connors, *The Race to the Intelligent State: Charting the Global Information Economy in the 21st Century*, Oxford: Capstone, 1997.
47. Prahalad, 'Strategies for Growth', p. 67.

12
Futures Studies and Co-evolutionary Futures

ANNE JENKINS AND MORGEN WITZEL

We need to confront the colonisation of the future by the Anglo-American West. The combined forces of Westernisation, commercialisation and technology dependency have produced a powerful dominating paradigm that makes this an urgent issue. At some point, we must bring non-Western and more peripheral futures into sharper focus; if this does not happen, then futures studies is doomed to an existence that will be, at best, sterile, and at worst, destructive.

The conflict, in our view, is not so much about West and non-West as it is between globalism and localism. The two conflicts are related but not necessarily interdependent. Globalism is about the creation of a single, world culture, based on the current ever-increasing expansion of Anglo-American culture, social norms and ethics and, last but not least, ways of doing business. Localism is about the survival of local identity, local cultures, local communities and all systems that sustain and nourish localism. Globalism seems to have a lock on both the present and the future—reducing both to a single denominator. Localism's survival may depend on its ability to generate its own futures and to force an acceptance of many, diverse futures on an unwilling establishment.

The best hope for futures may ironically come from the past. This does not mean that we are resurrecting an old heresy that the future can be predicted from the past. Rather, the disciplines of history and futures studies are similar; they use the same techniques, have the same responses to uncertainty and, indeed, fight the same battles for acceptance in a society that seems determined to ignore, or at least unwilling to confront both the diversities of our pasts and the pluralities of our futures. History teaches us about diversity. The various disciplines within history—economic, social, political, military, anthropological, scientific, feminist, cultural—provide us with our finest repository for understanding how diversity arises and what its consequences might be.

Recognising our own role as agents in the discontinuity between past and future provides us with further evidence of the need for diverse futures. Human agency creates diversity. The need to break away from the kind of linear thinking created by neo-classical economics, on which so much of current forecasting futures studies seems to be postulated, becomes of paramount importance. By recognising this we would not only provide space for other larger paradigms of futures—for example, Chinese, Indian and Islamic—to co-exist with those of the West, but we would also allow for personal models based on locale, communities, gender and occupation to flourish. The future belongs to everyone. If we allow it to be colonised by one dominant group, then we are in danger of surrendering our own futures to that group.

We will argue not only that it is possible for different paradigms—civilisational and others—to flower or mushroom in the contemporary world but also that it is an imperative for our very survival. Human cultures, like agriculture, need diversity to survive; monocultures are doomed to extinction. We maintain that there are strong similarities between the disciplines of history and futures studies; and an appreciation of history would not only open up futures studies but would also lead to individual and communal empowerment: human action in the present. Finally, we will argue that the concept of co-evolution[1] allows for the linkage of diverse yet harmonious cultural paradigms, in a manner that permits individuals to understand and, at least to some extent, influence and determine their own futures. Multiple paradigms, historiography and co-evolution provide us with a recipe to open up futures studies to multiculturalism and pluralistic possibilities.

Global Forces

For the past forty years and more, the prophets of globalisation have been predicting a kind of global convergence of cultures and structures. No one was more persuasive than Marshall McLuhan,[2] who foresaw the emergence of a 'global village' with communities tightly bound together by media and technology, which itself had a global reach. Though not all of McLuhan's predictions have come to pass, and though other important environmental forces have emerged, McLuhan's vision of a global community unified through technology has remained the dominant one, shaping the thinking of many planners, businesspeople and politicians across the world. It is a vision with prospects of becoming real. We are constantly dazzled by the power and speed of technology, by the prospect of doing things that could not be done last year, last month, last week. But technology, as ever, is only a tool. Visions of globalisation and unification of values go back much further than McLuhan in Western thinking. McLuhan himself drew heavily on the works of sociologists such as György Lukács and Franco Fortini earlier in the century. Such visions were very popular and came in both utopian (Aldous Huxley) and dystopian (George Orwell) forms. In the previous century, the British Empire reached its peak of world dominance, becoming in the process the most powerful global organisation the world has yet seen, basing its strength at least in part on an end-view of converting the 'uncivilised' world to the benefits of British Protestant civilisation; the 'white man's burden', however one may look at it, is a programme for the spread of a single civilisation. Before this the medieval Catholic Church regarded itself as the rightful ruler of the world, in a spiritual and often in a temporal sense; and the Church drew its modes of thinking in this respect from the Pax Romana established by the Roman Empire, which in turn drew its modes of thinking from the Hellenistic World of Alexander the Great and Athens.

Suddenly, we have gone from the modern age and the delights of technology to the foundations of the Western intellectual paradigm, the age of Socrates, Plato and Aristotle. There is no escaping this: as Alfred North Whitehead once remarked, we in the West are all intellectual descendants of Plato. Our intellectual and spiritual traditions are as impregnated with Platonism as those of China are with Confucianism. And one of the most important inheritances from that time is the concept of the dichotomous distinction: things are either one thing or the other, they cannot be both. Thus we speak of good and evil, right and wrong, black and white, light and dark, heaven and hell, capitalist and communist, democracy and dictatorship, war and peace. These are the terms by which we perceive the world around us.

This dichotomous perception is an integral part of the very necessary ethical framework within which we live our lives; and as the (perceived) alternative is anarchy, this is probably a good thing. The problem comes when we begin to apply these ideas in other settings. Then we begin to stray into the areas of intolerance and repression. Thus the medieval Catholic Church launched crusades against Islam; the British Empire (in some parts, at least) strove to stamp out local cultures and religions; the Anglo-American model of corporate management has striven, in the name of the free market, to overwhelm local business models and local market arrangements. Global brands, such as Coca-Cola, Sky Television and Nike running shoes, have succeeded at the expense of local companies and local products.

This is how the free market operates, if we accept the Western-based assumption that the free market is a good thing. And it can be argued that it is up to local firms to compete and respond. In some parts of the world, notably China, they are doing just that; Coca-Cola and McDonalds are already running up against stiff opposition from local competitors. More deleterious, however, is the spread of the Western business model into these same cultures. We are not only selling them our products, we are telling them how to run their businesses, and, in the long run, how to run their own lives.

The Anglo-American business model is rooted in two concepts: the separation of ownership and control, and the professionalisation and formalisation of management. Its legitimacy is virtually unquestioned. Its past has been validated; the works of Chandler[3] and others have 'proven' that this model of business developed naturally and spontaneously out of the rise of the modern corporation, which, coincidentally, took place in the United States. Its present dominance is unquestioned. Therefore, its future must be assured. Large management consulting firms, mostly of American origin, earn their billions by preaching the gospel of the American 'one best way' around the world; generic management tools and models are imposed on companies from Michigan to Manchuria. It goes without saying that any other way of doing things is wrong; in the words of one management consultant, 'Our way has a proven track record of success. We know it works. We can't see why anyone would want or need to do things any differently.'[4]

The gurus of the 'one best way' in business—Ted Levitt, John Naisbitt, Tom Peters—seek their legitimation beyond the world of business, however. They point to global trends, most notably the fracturing of the Soviet Union and the positioning of the United States as the one clear superpower, as vindication of their beliefs. Their intellectual icon is Francis Fukuyama.[5] The 'American way of life', as it used to be called, has been shown to be clearly superior to those of other lands. Other countries either follow and participate in this vision or are considered to be outside the pale. It is these latter countries that we in the West choose to lecture on democracy, human rights, personal freedom and the free market, concomitant values in the shining new future.

The Return of History

And yet, the signs are growing that this vision of unity will never come to be. At the same time as globalisation looks set to become a reality, local interests are also becoming stronger. Ironically, the same technological revolution that enabled global forces to spread far and wide is also enabling local communities and even individuals to increase their reach and voice. This was not foreseen by McLuhan or the utopian thinkers, and it is a concept strangely lacking in the writings of Fukuyama and other members of the 'end of history' school. If you give people personal liberty, what did you imagine they would do with it?

The European Union (EU) offers a classic case of localisation. On the one hand, the EU is oriented towards the creation of a 'superstate' in some form, a federation of nations that will be stronger than the sum of its parts, based on common economic and social factors. But even as the EU increases its strength and power, local groups within the EU's member states are flexing their own muscles. In the UK, the Scots and the Welsh have voted for devolution. In France, the Bretons and Basques are becoming ever more vocal minorities, and the long-submerged Provençal culture is reappearing. Italian regionalism offers a serious threat to the integrity of Italy as a nation-state. Even in Germany, minorities such as the Sorbs and Wends are striking out for their rights and demanding to be heard. America, the melting pot, has its own localisation problems. Canada teeters on the brink of fragmentation, and in the United States, the great expounder of globalisation, first black and then Hispanic minorities have risen to challenge the dominant classes of the nation.

These local pressures can be long submerged and forgotten by the rest of the world. The collapse of the federal states of, first, Yugoslavia and, second, the Soviet Union reminded us violently of ethnic groups whose existence we had almost forgotten. In the Americas, from Canada to Peru, native American groups are also re-emerging from their long, colonial history and finding their voices, fighting the dominant classes with the latter's own weapons; well-educated sharp-suited Haida lawyers argued their tribe's land claims in the courts of British Columbia, while in the mountains of Chiapas in southern Mexico, the leaders of the Frente Zapata use fax modems and Internet bulletin boards to distribute their message around the world.

What price, then, the end of history? As the 20th century draws to a close, history is staring us in the face. The long interlude of fascism and communism is over, and we are back to a world that looks perilously like that before 1914: fragmented, uncertain, heavily armed but with no clear enemies in sight, where competing interests jostle for power and influence. Where, then, are our futures? And where does this leave the futurist trying to understand them?

Multiple Paradigms

Acres of print have been devoted to the subject of paradigms ever since Thomas Kuhn first introduced the notion in his seminal work *The Structure of Scientific Revolutions*.[6] One of the key concepts about paradigms is that they change. Ways of thinking about the world and understanding undergo revision, until one day we realise that previous conceptions are no longer valid. Thus Einstein's physics overthrew classical physics; Galileo and Copernicus revolutionised astronomy; and now the theories of chaos and complexity are threatening to rewrite whole chapters of established 'scientific knowledge'. We know that all knowledge is tenuous, and that virtually any fact can be disproved. Yet we cling to our paradigms as articles of faith. We accept that there is 'one best way' and we blindly follow it, all forgetful of the idea that there might suddenly be another 'one best way'. More, we absolutely refuse to believe that there could be two, or more, paradigms ('two or more best ways') in operation at once.

And yet, as anthropologists and historians, at least, are beginning to demonstrate, each human being operates at any given time on one or more different levels. Both of the authors of this chapter are residents of England; both are citizens of the United Kingdom, which is quite a different thing, and both are resident in the European Union, which is a different thing again; both consider themselves to be part of a global community, which entails yet another set of rules and behaviours. Each is strongly attached to his or her region (northeast and southeast of England), community and so on. And this geographical hierarchy of belonging is only one dimension; we also have whole different structures based on ourselves as individuals, members of families, our roles in our workplaces, our academic interests and so on.

Is there conflict between these roles? Of course. Managing these roles in such a way as to create optimum convergence and manage role conflict means that each of us becomes (more or less) adept at managing paradoxes.[7] This is well recognised in postmodern theories, particularly in management, where authors such as Charles Handy[8] have been writing for years about the need to manage multiple roles on both a personal and an organisational basis, and Gareth Morgan has written equally cogently about the need to use multiple metaphors for organisation analysis.[9] In Morgan's view, using multiple paradigms or modes of thought simultaneously adds value. Referring to the inadequacies of perception, Morgan points out that we can layer different modes of thought on top of one another, each allowing us to see something that the others would miss, thus enhancing overall knowledge.

Diversity is natural, in society as well as in the natural world.[10] If its presence enhances knowledge, we can only conclude that its absence denies or detracts from knowledge. Neo-classical economics widely ignores Schumpeter and Veblen, for example, because their analysis of economic activity does not fit the dominant paradigm. The Western 'one best way' model of doing business, alluded to above, denigrates non-professional business structures such as the Chinese family business or Islamic systems of banking and finance because they do not conform to the model. By failing to place value on the characteristics valued by others, we risk not only the loss of knowledge entailed by a closed mind, but also the long slide into prejudice and repression. This is the challenge that futures studies must face, and this is why the colonisation of the future must be prevented.

Female Futures

Women, as Ho Chi Minh once said, are one-half of society, and their role needs to be recognised accordingly. Yet one has to look hard to find a feminine perspective in the dominant model of futures studies. This represents an exclusion even greater than that of other cultures or regions; effectively, half the population has been locked out of futures studies. Over the last thirty years or more, feminists have been prominent in seeking to highlight issues of power, privilege and dominance, exactly as embodied in the colonisation of the future. They have shown how academic work, for example, can contribute to social and cultural processes without simply reinforcing existing structural inequalities and demarcations. Feminist and gender frameworks are pressing for a rethinking of current categories and paradigms. Feminist research frequently challenges the common vantage points—the perspectives of civilisations, world-systems and international relations, for example—and draws on the viewpoints of more humble members of society such as workers, housewives and migrant peoples.[11] The body of knowledge that has been built up over this period of time has resulted in the creation of a distinctly feminist paradigm of knowledge.[12]

The locking-out of this paradigm from futures studies[13] has serious consequences. Even if we assume a globally dominant paradigm, this still means that half the population is defining futures for the other half. But, of course, if we add the feminist paradigm to cultural paradigms, then the discrimination becomes still more marked. Where, in futures studies, is there a consideration of the role of and future for women in Islamic societies? Where, in the brave new world of Chinese business, is there consideration of the future role of women managers, who have historically served ably and in large numbers in Chinese industry and enterprise? The feminist paradigm serves as a potent example, if example were needed, of the kind of damage that the colonisation of the future is doing to the discipline. Exactly the same sort of discrimination once existed in history, where, as one historian once remarked, women existed only as saints, scholars or slaves. Over the past several decades, however, the feminist paradigm has made strong inroads into the study of history, and it is now widely accepted; though much work remains to be done to bring 'women's history' on to a par with the rest of the discipline, the right road is being followed. This is just one example of the ways in which futures studies could usefully learn from history.

Futures and History

After some decades of experimentation, it is now fairly widely recognised that the past is not an accurate guide to the future. Attempts persist to use past behaviour to predict future actions (chartism, the predicting of future share price movements by past movements, is an example) but are not widespread. That raises the question of what the past can contribute, to which we will return later. For the moment, it is worth considering that of all the academic disciplines, the one with which futures studies has the most in common is history.

The study of history is, first and foremost, the study of uncertainty. We can never know what happened in the past; we can attempt to prove it beyond a reasonable doubt (there are strong parallels between history and law as disciplines as well), but there are always moments in time that remain dark to us. As we progress further back through history, those moments become more frequent and longer. Historians, therefore, become trained in the use of judgement, to make reasonable assumptions about those things that cannot be known. Part of this exercise of judgement entails the consideration of alternatives. Hard and fast conclusions, especially in more ancient history, are therefore rare; any piece of evidence is open to interpretation, and multiple conclusions are possible and common. Fisher illustrates another likeness history has for futures studies:

> Men wiser and more learned than I have discerned in history a plot, a rhythm, a predetermined pattern . . . I can see only one emergency following upon another as wave follows wave, only one great fact, only one safe rule for the historian: that he should recognise the play of the contingent and the unforeseen.[14]

Popper sees the idea that 'history depends in part on ourselves' as 'much more meaningful and noble than the idea that history has inherent laws'.[15]

Diversity at a micro scale is paralleled by diversity on a macro scale. The second point to be observed about history is that it is the study of diversity. History is full of diverse individuals, locales, belief systems and cultures. It is also dynamic; people and concepts emerge and die before the eyes of the historian. Nothing stays the same. Change and diversity are part of the links connecting the mosaic of evidence that the historian studies.

To deal with these things, historians build models. Briefly, one of the most powerful and persuasive macro models used by historians is that of the civilisation. The use of civilisations as a basic macro unit of historical analysis was pioneered in the 14th century by ibn Khaldun, the founder of sociology.[16] Ibn Khaldun's work was followed by Comte, Hegel, Sima Tan and Sorokin. In *A Study of History*, Arnold Toynbee[17] produced a topography of civilisations of the world from 3500BC to AD2000. Later generations of historians, such as Fernand Braudel,[18] further developed the idea of civilisations as 'ways of thought', that is, common systems of belief and knowledge that characterise the members of a society. Each civilisation thus effectively emerges as a paradigm for those people who live in it.

Civilisations—like people, like paradigms—rise, evolve and die. They have different roots, they go down different roads and they meet different ends. They also co-exist at the same time in different parts of the world. Although this conclusion is vociferously protested against by the proponents of the globalisation school,[19] there are a number of different civilisations co-existing in the world at present, of which the Anglo-American West is only one; India, China and Southeast Asia (Greater China, in some terminologies) and the Islamic nations are civilisations, as defined by ibn Khaldun, Toynbee and Braudel; and throughout their histories they have always seen themselves as civilisations.

None of the above is meant to suggest that history is a perfect discipline; on the contrary, historiography is fraught with prejudices, biases, intolerances and methodological imperfections, perhaps more than most disciplines (does this too raise an echo for the student of futures?). However, history is at the moment in the relatively happy position of having no one school or bias in a dominant position, and the student is free to pick and choose from competing interpretations. Thus there is diversity also in the way in which the study of history is conducted, which, to some extent, is leading to a greater tolerance of differing views.

Macrohistory also has something to teach futurists, in that throughout history there have been periods of both great cultural and technological diversity, and periods when one culture was dominant. For example, Alexander the Great's Hellenistic Empire swept away or engulfed many previously powerful civilisations such as those of Egypt and Persia; later the Roman Empire would do the same thing. The importance of a macrohistorical perspective in futures work is exemplified by Sohail Inayatullah.[20]

The Conformity of Fear

The Cambridge economist Joan Robinson once described the present as a lens through which both past and future are viewed; Rick Slaughter provides an excellent illustration of this.[21] Certainly our present actions can distort perceptions of both past and present, but we argue here that they do more; human action is the prime cause of the discontinuity (and indeed the dysfunction) between past and future. The power of human agency is of prime importance for both historians and futures studies, whether that be in terms of 'irrational' action that creates discontinuities or path-dependent actions that lead to conformity.

In *Mythical Past, Elusive Future*, Frank Füredi portrays a kind of reverse dystopia.[22] The future, he argues, has let people down. The old certainties are gone; even the dubious comfort of knowing who the enemy was, which existed during the Cold War, has vanished. Now we have the perception of being under threat, but we do not know who the enemy is. The various straw men erected by politicians from time to time—Moammar al-Qadafi, Saddam Hussein—serve as only temporary comforts for the peoples of the West. Science, which once promised us everything, now seems full of threat. Technology, enthusiastically adopted by some, is equally vehemently rejected by others, and this is creating another blurred vision; one person's utopia is another person's dystopia. (US Vice-President Al Gore's celebrated comment that people without access to the Internet do not have views worth hearing gives some idea of both the power of the paradigm and the range of people it excludes.) Fear of an uncertain future, Füredi goes on to say, is leading people towards more concentration on the past, but this is unrealistic behaviour; people are turning to the past for comfort, nostalgia, for traditions to cling to, rather than for real learning.

The late 20th century is a world full of fears, the most prominent of which is the fear of uncertainty. Our perceptions suddenly seem blunted and the very things that extend our knowledge—technology, communications—bring us up against the seemingly impenetrable wall of what we do not know, and what we may never know. As a defence against uncertainty, we close ranks around what is known, around those things that are certain.

One argument for globalisation is that the reduction of diversity means a reduction of the unknown. Proponents of the single European currency, for example, argue that currency union will eliminate currency fluctuations and therefore one level of business uncertainty. This may be true, but it confuses diversity and lack of knowledge. It is diversity that is *unknown* that leads to uncertainty and fear. If we know about diversity, we can confront it, learn from it, accept it, oppose it or adopt whatever strategy seems best. Although globalisation purports to reduce uncertainty, in fact its real effect is merely to shut it out. The problem of confronting the future is postponed, not eliminated; there is no end of history, merely a caesura. Worse, as already mentioned, refusing to interact with other paradigms, civilisations, call them what you will, creates ignorance. This is the kind of ignorance that led the West to denigrate Islamic and Oriental societies as being inferior, and to refuse to recognise the value that could be drawn from them.[23] The combination of ignorance and fear is exactly what we have seen during this century, which, despite all our advances, is considered one of most violent centuries in history, producing fascism, two world wars, countless other wars and genocide on a mass scale.

The human desire for universal truths is fundamental—it can be found across all cultures; in this respect, at least, Plato was not different from Confucius. Its most common form today is the Great Universal Search for Everything in which physicists are presently engaged.[24] Perhaps this is the right paradigm for physics; time will tell. But in other areas, perhaps we need to start rethinking what we mean by 'universal truths'.

Different Pasts, Different Futures

Diversity, we argue, is itself a universal truth. It is validated by history. We know now that there is no single, all-pervasive civilisation—human cultures and societies emerged at many different points on the globe and developed in ways that exhibited at least as much diversity as similarity. Without getting involved in the tricky and complex debate about the relationship between diversity and similarity, we feel we can state without fear of contradiction that diversity is as much a part of the human inheritance—genetic as well as social and cultural—as similarity.

A relationship between past, present and futures may be of help in breaking the present cultural lock on this field. The past is full of diversity, and that diversity is reflected in the present. At this point, thanks to the discontinuity of human actions, we have two choices. We can either remain dependent on what we see as the path laid out for us by the past and proceed on that basis into the future; or we can break with the past, and set ourselves on a new course. Individuals, communities, states and civilisations all have this power to determine, at least to some extent, their behaviour at a given point in time. However, their behaviour is strongly constrained by their lack of knowledge of the future. They do not know the consequences of their own actions, nor do they know what environmental trends might impact on them.

Futures studies can help reduce this uncertainty by presenting people with a picture of how the future might look. Uncertainty is not thereby eliminated, but it is reduced to some degree. People are able to plan with varying degrees of confidence, make contingencies in case the plans do not mature, and so on. This is precisely the role played by forecasting in modern business organisations. However, in most cases this is where it stops. The feedback loops are not capable of incorporating learning about the future into learning about the present or past. A diversity of futures allows different futures to be shaped by different individuals according to their context. To use the example cited above, a world future, a European future, a British future, an English future and a local future are all perfectly possible. Nor do they nest inside each other, like Russian matrouschka dolls; they can be in serious conflict. It is there that the art of managing paradoxes becomes important—and it is on this art that we believe the emphasis must lie in future education and training.

Rather than simply increasing diversity, the offering of diverse futures will actually reduce uncertainty. Futures are, after all, no more than packages of knowledge.

Just as with Gareth Morgan's organisational futures, a cluster of futures will offer more knowledge than a single future. Following the logic that the possession and use of knowledge reduces uncertainty, we believe that diverse futures will make people less uncertain and, assuming they have the skills to manage paradoxes, plan for the future with more assurance. With greater assurance of the future will come more respect for the past. People will begin to see the past not as a source of crutches to prop them up against present insecurities, but as a well of inspiration for their own actions and behaviour in diverse situations. By learning better how to cope in the present, they will become still more tolerant of uncertainty in the future and better able to act and respond. The process becomes circular and self-fulfilling.

There is obviously a long way to go before such a picture can be completed, and the need for education is certainly a major stumbling block. For a start, however, we need to focus on futures studies itself. Unless futures studies can break the stranglehold of the dominant paradigm and open the door to diversity, then the entire effort would seem doomed.

Co-evolving Futures

How can diversity be integrated into futures studies? This is the task that now confronts futurists all around the world, and to which their attention must now be given.

Of foremost importance is the need to *promote* rather than *preserve* diversity. We have indicated the need to integrate history with futures studies, but we must not become overdependent on the past. Preserving our identities and values means letting them emanate from our cultures; it does not mean returning to a fossilised past, romanticising tradition and banning Western media and access to the Internet (as happens in Iran and, to a lesser extent, in China) in an attempt to preserve this tradition and prevent 'Westernisation' or 'deterioration' of a culture. This is akin to trying to hold back the waves, and it is doomed to failure. The reality is that cultures and civilisations are dynamic and in a state of continuous change. If we attempt to freeze cultures in time, then they can no longer serve as guides to people in their current environments; their meaning and legitimacy is rightly called into question, as indeed is happening in parts of the 'heritage' industry in Britain today.

The second factor we need to consider is the extent to which different cultures, different paradigms are linked. If all are in a state of dynamic evolution, then they are equally linked and in effect co-evolving, linked to and influencing each other. Thanks to modern technology and media, those links are more powerful and faster acting than ever before. Cultures evolve, in part at least, through interaction with each other. But the surface picture can be very one-sided. We see that the West 'loots' other cultures by adopting and adapting selected features; the Americanisation of Japanese foods such as sushi in the 1980s is an example of this, as too is the emergence of so-called world music around the same time. Alternatively, Western civilisation invades other cultures, penetrates the deep enclaves of non-Western societies with its consumer products, and leaves a strong imprint of consumer culture on them—even a tribe of San aborigines in the Kalahari desert has now reportedly acquired a taste for Coca-Cola, complete with Coca-Cola T-shirts. But, this does *not* mean that we are heading towards a global culture. Such a thing exists largely in our own perceptions. Other cultures are adept at scanning and borrowing techniques and cultural attributes from the West without making wholesale changes to their own way of life. Deng Xiaoping once remarked that the way to reform China was to study the West, borrow those things that worked, discard those that did not, and integrate the former into the Chinese way of doing things; this philosophy has been and is practised in China, with notable success. One of our greatest mistakes in the West is failing to learn from other cultures in a similar way. We have mistaken their demand for our knowledge for an indication that our knowledge is superior, and we denigrate their own knowledge accordingly.

Co-evolution does not and will not result in homogenisation. What it does is integrate temporal diversity with cultural diversity. We use this concept to describe not only the impacts of time on cultural paradigms but also the impacts of space and location on historical and future paradigms. We cannot, and do not, live in isolation, either temporally or spatially. Tolerance of diversity and change must be respected and promoted. We can, and must, live according to Kant's words, 'Dare to be free and respect the freedom and autonomy of others'. We can, and must, integrate co-operation and competition, in the fashion described by Axelrod.[25]

The challenge for us as world citizens is to do our best to make futures meaningful.[26] In doing so, they must be meaningful to individuals. People want and need to own their own futures; the futures of others are useless to them. Our final plea is to give futures meanings to individual selves, to engage people in their own futures and make the future as meaningful as the past. For, if we do not engage in our futures, then we risk having control of our futures taken away from us. In effect, we risk losing our identity to the whims and decrees of others. Here is the gravest danger of all posed by the colonisation of the future: that we will be left with a future in which individualism has no meaning.

Futures studies exists because of uncertainty about what is going to happen next. By providing familiarity with possible outcomes, futures studies attempts to reduce uncertainty. But more than that, it gives people back their power to influence their own futures and change and shape their own identities and lives according to their own aspirations.

Notes

1. Stuart A. Kauffman, *The Origins of Order*, New York: Oxford University Press, 1993.
2. Marshall McLuhan, *The Global Village*, New York: McGraw-Hill, 1998.

3. Alfred D. Chandler, Jr., *The Visible Hand*, Cambridge, MA: Harvard University Press, 1977.
4. Conversation with author, Beijing, October 1996.
5. Francis Fukuyama, *The End of History and the Last Man*, London: Hamish Hamilton, 1992.
6. Thomas Kuhn, *The Structure of Scientific Revolutions*, Chicago, IL: University of Chicago Press, 1962.
7. Jagdish Sheth, *Managing Your Self*, New York: Wiley, 1992.
8. Charles Handy, *Understanding Organizations*, London: Penguin, 1993.
9. Gareth Morgan, *Images of Organization*, Newbury Park, CA: Sage, 1986.
10. David Riesman, *Thorstein Veblen: A Critical Interpretation*, New York: Seabury Press, 1960.
11. L. Morley and V. Walsh, *Feminist Academics: Creative Agents for Change*, London: Taylor & Francis, 1995.
12. Readers may object to our use of the word 'paradigm' with different meanings in different places. To this, we respond that Thomas Kuhn used the word 'paradigm' with no less than thirty-two different meanings during the course of *The Structure of Scientific Revolutions*, a fact that he acknowledged himself in his later writings.
13. Ivana Milojevic, 'History, Feminism and Futures', *Futures* 28(6–7): 629, August–September 1996.
14. H. A. L. Fisher, *History of Europe*, 1936, quoted in Karl Popper, *In Search of a Better World*, London: Routledge, 1994.
15. Popper, *In Search of a Better World*.
16. *The Muqaddimah: An Introduction to History*, translated by Franz Rosenthal, Princeton, NJ: Princeton University Press, 1967.
17. Arnold Toynbee, *A Study of History*, London: Oxford University Press, 1960.
18. Fernand Braudel, *A History of Civilizations*, Harmondsworth: Penguin, 1987.
19. See e.g. John Naisbitt, *Megatrends*, New York: Warner Books, 1984.
20. Johan Galtung and Sohail Inayatullah, *Macrohistory and Macrohistorians*, Westport, CT: Praeger, 1997.
21. Richard Slaughter, *The Foresight Principle*, London: Adamantine, 1995.
22. Frank Füredi, *Mythical Past, Elusive Future*, London: Pluto Press, 1992.
23. Edward Said, *Orientalism*, London: Routledge & Kegan Paul, 1980; Max Weber, *The Protestant Ethic and the Spirit of Capitalism*, London: George Allen & Unwin, 1930.
24. Stephen Hawking, *A Brief History of Time*, London: Bantam, 1988.
25. Robert Axelrod, *The Evolution of Co-operation*, London: Penguin, 1984.
26. Karl Popper, *In Search of a Better World*, London: Routledge, 1994.

13
Futures Studies and the Future of Science

STEVE FULLER

Science as the Open Society: An Elusive Ideal

In *The End of History and the Last Man*,[1] Francis Fukuyama declared that a liberal democratic future awaited all the world's peoples, courtesy of capitalism's systemic beneficence. Fukuyama was only the latest of a long line of thinkers who claimed to have divined the future of humanity by extrapolating from current trends. However, when it came to explaining how we shall have realised our destiny, Fukuyama was one with his Marxist opponents in pointing to the 'logic of natural science' as plotting an inevitable course that both transcends and transforms even the most historically entrenched of cultural differences. In that sense, science puts an end to history: once the natural trajectory of science is appropriately harnessed to the future of one society, history then simply consists of the rest of the world catching up by repeating the steps originally taken by that society. Until quite recently, this was how both capitalists and socialists in the first two 'Worlds' thought that the Third World would be 'modernised'. Socialists pointed to science's role in the creation of labour-saving technologies that eventually undermine the basis for any sharp distinction between the workers and their bosses. For their part, capitalists emphasised the role of science in enhancing people's innovative capacities and hence their ability to compete more effectively in the market-place. The roles assigned to science in the two political economies were different, but both were meant to have globally liberating consequences.

Their influence on policy notwithstanding, theorists of global development and modernisation generally presuppose rather unsophisticated, uncritical accounts of the nature of science. A common feature of their accounts is that science is unique among human institutions in its long-term progressive character. Moreover, as we have just seen, the presumed progressiveness of science can be used to justify the progressive character of any number of contradictory political and economic regimes. Indeed, sometimes it seems that 'science' is little more than the name given to whatever progress the history of politics or economics is said to display. Thomas Kuhn had a characteristically equivocal way of capturing such accounts of science that straddle between describing its actual conduct and the standard it sets for the rest of society.[2] Kuhn professed an interest in accounting for science when 'it functions as it should'. Tactfully omitted from this aspiration was any judgement about how often, if ever, science lives up to its own standards of rationality and objectivity, the ones that then provide the normative basis for today's so-called knowledge societies.[3] Consequently, Kuhn, like so many other theorists of science, suspends his account of science in what might be called a state of *artificial positivity*, that is, the dubious belief that the clarity with which 'science' can be articulated as a normative ideal is indicative of the ideal's realisability in today's world. Let me briefly explain this concept, as it will provide some background understanding to the problematic status of science in most policy studies, especially future studies.

'Artificial positivity' is modelled on 'artificial negativity', an expression associated with the arch-scepticism of the Frankfurt School's version of critical theory, which saw capitalist complicity in every form of cultural production. Such conspiracy theorising, albeit conducted at a very high level of abstraction, ends up winning intellectual battles while losing the political war, since it induces the critic's withdrawal from the public sphere, lest the critic be sullied by capitalist conspirators. 'Artificial positivity' represents the complementary attitude that there is nothing for the critic to do because the ideal is already presupposed in everyday practice, sometimes a 'transcendental precondition' of such practice, as Habermas might put it. In that case, any perceived discrepancies between the ideal and the real are treated as localised incidents, the remediation of which will occur in the long run, either because the system naturally corrects itself or because people come to see the discrepancies as systemic virtues in disguise. Normative visions of science as the 'open society' are typically subject to artificial positivity in this sense, with the result being a bland endorsement of the status quo. For example, the fact that many cases of research fraud are eventually caught by scientists themselves is taken to vindicate the self-critical function of science, not to signify a deeper, more systemic problem with the conduct of scientific research. Also, the fact that science displays a pecking order of researchers, institutions and even subject areas that rivals that of any class-based society is presumed to be the desired outcome of processes involving the free and open participation of all members of the scientific community. The fact that these processes cannot be easily specified and that many scientists are clearly dissatisfied with their place in the pecking order are treated as areas 'in need of further empirical investigation', not indirect proof of the artificially positive assumptions made about the realisability of the open society in science today. Perhaps the most thorough recent defence of a vision of science in this artificially positive mode is Stephen Cole's *Making Science*.[4]

In light of the above, it is no surprise that the arch-sceptics of our own time, the postmodernists, have rejected the open society ideal of science as just so much wish-fulfilment. Nevertheless, the ideal remains sufficiently alive in policy circles and is sufficiently admirable on its own terms to deserve a rearticulation, one that deals squarely with the political and economic conditions that are necessary for its realisation. This project goes very much to the heart of my own programme of *social epistemology*. When I began this programme, just over ten years ago, I wanted to lay the foundations of a sort of welfare economics of science, or 'knowledge policy'.[5] While this still captures my general normative sensibility, the political implications of my work have moved between liberalism and socialism, depending roughly on whether I have drawn my disciplinary resources from the humanities or the social sciences, respectively. However, implicit throughout has been a commitment to the values associated with what Karl Popper popularised as 'the open society' and political theorists call 'republican' regimes.[6] Readers with well-tuned world-historic sensibilities may find this admission tantamount to an endorsement of the most backward-looking Eurocentrism. After all, the list of republican regimes typically includes the West's iconic political settings: classical Athens, pre-imperial Rome, the city-states of Renaissance Northern Italy, Whig Britain, and the US Constitution's rule by 'division of powers' and 'checks and balances'. However, I suggest we start by asking why it has been so difficult to establish republicanism on a long-term and worldwide basis. These difficulties revolve around the need for a society to provide the economic security and cultural resources needed to protect its members' *right to be wrong*, which at a more abstract and collective level amounts to a policy of preventing the past from overdetermining the future. I then focus these considerations on the university, which, since first acquiring self-governing status in Roman corporate law, has been the institution most indicative of the tensions inherent in carrying out the republican project. Finally, I turn to the implications that this recovery of republicanism has for future studies, which has yet to engage with the past to the extent that I suggest here is necessary for forging an empowering future.

Situating Republicanism between Liberalism and Communitarianism

Most of the debilitating effects of political and scientific regimes have resulted from people feeling that they cannot either admit their own errors or reveal the errors of others—unless the errors are minor ones, of course. If you find 'error' a bit too pre-postmodern, how about the capacity to change one's own and others' minds in public? I happily grant that those who propose claims about the errors of others may themselves be in error. However, for most of history (including the present), people have been afraid even to *speak* in terms of their own or others' 'errors' because of what they fear to be the consequences of such talk. The result is that a self-imposed authoritarianism can remain in force even in avowedly 'liberal' and 'communitarian' societies: the former find the prospect of errors too risky to bear individually, whereas the latter portray them as the betrayal of duty to the collective.

In what follows, I must plead guilty to the charge of conflating the political and scientific sides of the republican ideal. My reason is that, strictly speaking, science is a representative body, whereby a few speak for the many. If this were not literally true, then what exactly would science's claim to 'universal knowledge' mean? Yet, scientists are not elected by the populace, or even a representative sample of the populace. Rather, they are 'self-selected', which means that people who are already scientists—and then relatively few of those—decide who is fit to hold the title of 'scientist'. The mystery surrounding science as a political concept has less to do with how it conducts its day-to-day business (i.e. 'research') than with how it comes to speak on behalf of the whole of humanity: specifically, science's ability to provide global governance of the human condition in a way that transcends national differences as well as other cultural and economic barriers. On the one hand, the public trusts—or at least defers to—scientists, though few non-scientists have ever witnessed how scientists come up with the knowledge on which their judgements and explanations are based, and not many more can recite the catechism presented in science textbooks. On the other hand, the phenomena that scientists are authorised to judge and explain on the public's behalf typically have been experienced more directly by the public than by the scientists themselves (e.g. biological explanations of childbirth, drug-taking, etc.; psychological explanations of virtually anything). Very few, if any, religions have commanded such blind loyalty.

Republicanism's underlying idea is that freedom requires the expression, not merely the toleration, of different opinions. The realisation of this idea presupposes certain conditions: (1) that people's opinions might change for the better as a result of hearing opposing opinions; (2) that people need not fear the consequences of their expressed opinions on their material well-being; (3) that there is a 'public good' or 'civic ideal' to which people may appeal in deliberation that transcends specific individual and group interests. Republican policies aim to ensure that all citizens are sufficiently secure in their material circumstances that they are not inhibited from speaking their minds. If you can express your mind with impunity, then your ideas can die in your stead, to recall a phrase of Goethe's that Popper liked to use to epitomise the open society. The significance of this capacity should not be underestimated. A frequently remarked obstacle to instituting 'deliberative democracy' schemes (e.g. citizens juries, electronic town meetings) is the tendency for people to reinvent patterns of deference even in arrangements that have been designed as much as possible to be egalitarian.[7] While some trace this tendency to the inherently hierarchical nature of human beings (especially when large populations make it 'efficient' to distinguish sharply between leaders and followers), more likely it has to do with the fear of humiliation that comes from making one's mistakes in public.[8]

The political psychology surrounding the inhibition of republican sentiments is complex. While a measure of economic security is required for the realisation of the republican ideal, it should not be assumed that the poorest members of society are the most easily inhibited from expressing dissent. On the contrary, it may be argued that the inveterate poor have little to lose and hence are more willing to speak their minds than those who regard the middle class as within striking distance. Academia is one social environment that perpetuates this sense of bourgeois self-containment. Despite the clarity with which its 'haves' and 'have nots' are marked in terms of funding, publications and institutional location, very few academics stuck at the bottom of the pecking order ever believe that they are consigned to the dustbin of history.[9] Consequently, even the lowliest contract researchers believe that they are still contenders for tenured posts and hence think they potentially have something to gain by biting their tongues in deliberative settings. In this case, continued faith in an ideology of meritocracy is sustained by the vagueness of criteria for success and failure, combined with the smattering of success stories that compare slightly favourably with winners of the national lottery. If one wanted to make a case for the irrationality of individuals who opt for the pure pursuit of knowledge in our times, then this would be the place to begin.

The historic virtue of republicanism is its concerted efforts to mitigate, if not outright eliminate, most of the hereditary bases for wealth and power that have all too frequently overdetermined any given generation's level of achievement.[10] In past republican regimes, people who simply lived off their inheritance were despised and, when possible, dispossessed of their holdings through the levy of a heavy inheritance tax. The beneficiaries of this redistributed wealth were those who were likely to increase the wealth of all in the future (e.g. entrepreneurs) or had already prevented it from decline in the past (e.g. military and civil service pensions). Property ownership, typically a requirement for full citizenship, reflected less a deference to wealth as such than a basic political competency test: how can people be trusted to exercise independent judgement in the forum, and potentially offer their leadership to the entire polis, if they cannot even manage their own affairs? At the same time, the dispossession of inherited wealth was never total, since the market had yet to govern all forms of social interaction. The classical Greek sense of economy as *oikonomikos* still ruled, and so even the most inefficient and unproductive amongst the wealthy were left enough to maintain a household with some dignity. In that important sense, republicanism was 'pre-liberal'.[11]

A contemporary proposal in the same spirit is the 'guaranteed minimum income', which reflects that now, after 200 years of liberalism, there is a greater need to protect the poor than the rich from indignity. Moreover, wage labour has gone the way of land ownership in failing to capture the economic preconditions for making a meaningful contribution to society. In general, the value that republican regimes have invested in powers of self-maintenance has related to the source of such regimes' leaders, who may well be chosen by lot, as in the case of classical Athens. No doubt, citizens of our own 'democratic' societies would be gladly taxed much more heavily for improving education, if their leaders were selected in this manner. To his credit, John Stuart Mill had figured this out when he originally called for publicly funded mass education coupled with the use of education as the main criterion for political participation.

Reproducing republicanism has proven elusive, mainly because its identity is so closely tied to its catalytic role in the West's acceptance of the capitalist way of life. Philosophical histories of European politics present republicanism as a transitional phase between the close-knit communitarianism of agricultural societies and the dispersed liberalism of commercial societies.[12] There is a comparable philosophical history of science, whereby the ideal of the open society appears briefly realised in the Enlightenment's 'republic of letters' between the clerical authoritarianism of the feudal era and the privatisation of intellectual property rights that began in the 18th century with the institution of copyright and accelerated in the 19th and 20th centuries with the expansion of patent law. According to the tacit conventions by which history is reified as 'theory', communitarianism and liberalism—or normal science and technological innovation, respectively—are regarded as 'pure types' of governance, while republicanism is taken to be an unstable 'hybrid' (*especially* in the biological sense that hybrids are infertile). For, no sooner had republican regimes eradicated the privileges enjoyed by the Church and the landed aristocracy than social instability dissolved republicanism's own civic ideal. In its place emerged, over a couple of generations in the 19th century, a new industrial elite nurtured by laws permitting the transmission of acquired wealth to offspring, as justified by a liberal's new-found sense of individualism: namely, 'I am entitled to dispose of my hard-earned wealth as I please.' These legal arrangements begat the great dynasties and monopolies, perhaps most notoriously symbolised by the Rockefellers and Standard Oil. In response, governments had to invent new regulatory powers for themselves that either assimilated industry into the state apparatus (the continental European route) or cast the state in the role of 'trust-buster' (the US route). Both routes were designed to recapture, however artificially, the lost world of republicanism's civic ideal.

(*a*) *The scientific matrix*

	Public good	
Risk-seeking	Yes	No
Yes	Popperian bold conjecturer	Schumpeterian creative destroyer
No	Kuhnian normal scientist	Cartesian cogito

(*b*) *The political matrix*

	Civic ideal	
Risk-Seeking	Yes	No
Yes	Republican	Liberal
No	Communitarian	Robinson Crusoe

Figure 1. The scientific and political matrices

My argument so far can be summarised in the pair of matrices in Figure 1, which pursues the analogy between the organisation of scientific and political life.

Let me elaborate the two dimensions that define the terms of the analogy:

1. *The presence of a civic ideal/public good = Is there a commonly recognised 'court of appeal' that is transcendent of, and irreducible to, special interests?* Indeed, both special interests and private property presuppose the inviolability of a CI/PG such that its subversion would be self-defeating. Special interests are never favoured in a large polity for their own sake, but only to the extent that they can serve the interests of others. Private property is not *sui generis* but the product of a transformed commons. Intellectual property may be regarded in similar terms, namely as the application of universally available principles, or the economist's sense of 'public good'—a 'public good' in the sense that it costs more to restrict access to the good than to keep it freely available. In this way, economists have distinguished 'pure' from 'applied' knowledge, the latter restricted by intellectual property laws. Another example of the presupposition of a commons in science is that you cannot oppose a scientific theory without abiding by the same rules of evidence and method that are allowed to your opponent; to do otherwise would be to opt out of the scientific field altogether. If you want to alter these rules, you must propose the changes openly, which then makes them subject to public scrutiny. The presence of CI/PG provides an external boundary (or 'demarcation') to the political/scientific enterprise (say, by nation, discipline), which enables its internal changes to be clearly tracked and reckoned. At any given point, they constitute what Popper called 'conventions' and Kuhn a 'paradigm', which are recognised by the republican as *necessary evils*. (The mistake made by communitarians like Kuhn is to regard the evil as a good, whereas the mistake made by liberals is to suppose that this 'evil' is eliminable without negative consequences.)

2. *The degree of risk-seeking = Once the boundary of the polity/science has been set, are there incentives to change its internal constitution?* Where no such boundary exists—i.e. in societies governed merely by the dictates of individual self-interest—one gets radically polarised responses. Those inclined to risk their lives for their ideas will appear as heroic figures, whereas those who do not are seen as self-sufficient, not a burden on anyone else. For their part, republican societies have tended to use military conscription as the link between defending the polity from external threat and enabling citizens to fortify themselves in case their own lives become imperilled by the claims they express in public. This Italian innovation is captured by the second amendment to the US Constitution: the individual's right to bear arms.[13] While decrying the American tendency to vigilantism, I nevertheless believe that there is merit in the idea that tools honed to defend the polity can then be used to encourage the contestation of issues within the polity. Of course, we need to find a less violent means of transmitting the relevant skills and attendant attitudes. Indeed, compulsory citizen education of the sort advocated by J. S. Mill in the UK and John Dewey in the USA may be seen as sublimating

the republican society's interest in cultivating the martial arts. This is perhaps most evident in the rhetoric surrounding science education, which suggests that students are training to enable the nation to be more competitive in the global economic arena, while enhancing their own employment prospects at home.

First Republican Strategy: Shoring up the Commons by Taxing Social Inheritance

The most obvious difference between republicanism and liberalism is that the liberal does not recognise any sense of collective interest beyond aggregated self-interest; hence, liberalism has found the idea of a 'civic ideal' elusive, if not completely chimerical. This is because societies dominated by the market mentality —as liberalism invariably is—make it rational for individuals to measure what they say against its likely consequences for their own well-being, assuming no social buffer from the repercussions of having made claims that are ultimately deemed mistaken. Thus, democratic political theorists sometimes draw a distinction between 'liberal' and 'republican' regimes in terms of the types of freedom their citizens enjoy. In liberal regimes, it is supposed that sheer lack of physical interference constitutes freedom. In other words, if I am not enslaved, I must therefore be free to do what I want: *tertium non datur*. However, republican regimes do not accept the premiss as sufficient to warrant the conclusion. In addition, republicans hold that people need to act in an environment where there is a good chance that what they say and do will be taken seriously by others, and not simply ignored or made the grounds for the curtailment of their speech and action in the future. In science today, *at most* its elite members live in a republican regime, while the rest live in a liberal regime where their freedom is in practice severely constrained by whether they will offend a prospective employer or grant reviewer.

I say 'at most' because the symbolic capital accumulated by scientists has come to be so bound up with ordinary economic capital (i.e. better = richer) that even dissenting members of the elite can have their right to be wrong seriously threatened by the 'liberalisation' of inquiry. One such elite researcher is Peter Duesberg, Professor of Cell Biology at Berkeley, who was recently stripped of his Outstanding Researcher status at the US National Institute of Health (which had virtually given him a blank cheque for research) for having publicised his scepticism that AIDS is caused by HIV, arguing that AIDS may be a straightforward public hygiene problem related to the lifestyle of gay men that then leads to the breakdown of their immune systems. Serious consideration of his hypothesis would clearly challenge the American medical establishment's backing of a strong HIV–AIDS link, while rekindling an unwanted public debate on the ethics of homosexuality. Of course, in keeping with US First Amendment rights to 'free speech', Duesberg did not lose his job or get thrown in jail, but his research funding and book contract were withdrawn, making it difficult for him to develop and advocate his position effectively. In contrast, an elite dissenter who has successfully adapted to the erosion of republican science is John Bockris, Professor of Chemistry at Texas A & M University, an early and continuous supporter of cold fusion as an important key to alternative energy research. Though widely criticised as high-tech alchemy, his lavishly and privately funded research has insulated him from calls by colleagues to divest him of his professorship. However, it is clear that were Bockris to discover a commercially viable cold fusion process, it would be immediately patented as a technology and hence be taken out of the domain of pure inquiry. The point here is that the motives for long-term investment in wild ideas should not be confused with the pursuit of knowledge for its own sake.

Fear and anxiety infiltrate the day-to-day operations of science, that is, even when scientists are not proposing controversial hypotheses. This fact is obscured by the norm that Robert Merton euphemistically labelled 'communism',[14] which in practice means that scientists have *no choice but* to share data and credit if they expect to be supported in the future. On closer inspection, a mafia mentality turns out to be at work, itself another by-product of the liberalisation of science. Consequently, scientists must worry about how their interests might be affected by whatever they say in public. Consider the peer review processes that govern scientific publication. They essentially provide insurance against risk: an individual scientist is allowed to say only as much as her peers can tolerate, and in return they absorb collective responsibility, so that she does not have to bear the burden of proof alone. Readers can simply rely on the judgement of the journal's editors to vouch for the veracity of the author. However, this insurance is purchased at a cost, namely, that the contestation of already existing claims to knowledge is kept to a minimum and, wherever possible, pre-empted by gestures to portray one's work as cultivating a domain charted by previous researchers. After all, a forum that publicised knowledge claims only to have them seriously challenged in subsequent issues would appear to be promoting more a clash of ideologies than the collective search for truth. In that sense, science is a society designed to suppress conflict through fear and anxiety rather than resolve it through either a peaceful consensus or open warfare.

In contrast, 'republican science' is inquiry in a state of 'permanent revolution', an expression that Popper[15] originally adapted from Trotsky to distinguish his position from Kuhn's[16] more sanguine attitude to normal science. To be sure, the expression is an oxymoron, albeit a suggestive one. If people are always encouraged to be critical of their current situation and to propose better alternatives (rather than simply wait until the status quo fails on its own accord, as Kuhn suggested), then revolution loses its violent character, mainly because people come to realise just how arbitrary the status quo really is—perhaps little more than an historically entrenched accident. In that case, they will be less inclined to superstitiously associate longevity with destiny. They will realise that there is just as much risk in continuing the status quo as in breaking with it, at which point the past comes to be treated more as a contingent resource than as a necessary burden or even legacy.

In economic terms, the permanent revolutionary is thus more concerned with *opportunity costs* (how many possible futures are closed off by my action?) than *sunk costs* (how many possible pasts were closed off by the actions of my predecessors?). Under these circumstances, induction and its rhetorical twin, presumption, lose their force as principles of rational inference. Similarly, at the level of economics, republicans treat initial investments into a particular line of inquiry as a 'sunk cost' whose continuation must be gauged on the likelihood of future return and not merely on the sheer irretrievability of what was initially invested. Such a strategy would immunise science-policy-makers against the belief in self-fulfilling prophecies and, still worse, the idea that wrongs sufficiently compounded over time add up to a right.

A republican science would treat what Robert Merton, in another of his classic euphemisms, called the *principle of cumulative advantage* as comparable to the untaxed inheritance of acquired wealth.[17] Merton originally had in mind the selective advantage afforded to scientists trained at elite institutions, considered in relation to their ability to accumulate grants, publications, posts and honours. Merton's followers have raised this tendency to a kind of 'invisible hand' at work in science. Since the distinction between those at the cutting edge of research and the rest of the pack is usually not the result of any central planning board, it must be the result of the collective spontaneous judgements of scientists themselves: so say the Mertonians. Neglected here, of course, is the role that the sheer size of the initial investment into a line of research (in terms of both labour and capital) plays in giving it forward momentum. Thus, the quick-and-dirty indicator of scientific progress in our times, its irreversibility (i.e. the inconceivability of turning back on an inquiry once begun), is measured more in terms of the amount that has so far been invested than the benefits that have so far been realised. From that standpoint, the US Congress's decision to halt the construction of the world's largest particle accelerator, the Superconducting Supercollider, in 1992 struck a blow for republican science against the superstition of irreversible progress. But generally speaking, what is the antidote to this pernicious mentality?

My first piece of advice is to regard the predictability that an Ivy League or Oxbridge degree will bring scientific success in the same spirit as Marxist and institutionalist economists have regarded the stability of the major corporate dynasties or oligopolies in capitalism—that is, as symptomatic of inequities in the way science has come to be institutionalised, not a sign that science has managed to sort out the wheat from the chaff. The appeal of both 'big science' and 'big business' is grounded in a myth of continuity, be it of a research programme or a corporate track record. In the case of the former, one speaks of the regularity with which such a programme can come up with novel findings; in the latter, the corporation's experience in manufacturing or, more likely these days, marketing new products. To political economists such as Thorstein Veblen and Joseph Schumpeter, who, in their different ways, placed a greater value on the spirit of enterprise—that is, the periodic destabilisation of stagnant markets—than on the sheer accumulation of wealth, the illusion of a 'weight of the past' sustained by lax redistributionist policies appears more as millstone than milestone to the continued existence of capitalism. Basically, capitalism's competitive edge is gradually eroded, as governments go for the soft option of allowing the perpetuation of wealth in ways that resemble its current form.

Here history provides a valuable lesson. In the early days of capitalism, before high productivity was realised, inheritance taxes were seen as the only equitable policy for redistributing wealth so as to ensure that markets remained free to all enterprising individuals. However, with the advent of high productivity came the modern welfare argument for taxation, which assumes the possibility of benefiting the poor without disadvantaging the rich. Thus, the rich would be taxed mainly for what the state regards as surplus income that could be used more efficiently for the overall benefit of society by being given to the poor. Arguments then revolve around the relative efficiency of taxing the rich in this way *vis-à-vis* providing them with incentives to invest their surplus income. But of course, seen in historical perspective, this debate amounts to little more than cosmetic accounting, whereby income levels are narrowed but the difference in relative advantage that the poor and rich receive from each unit of income remains unchanged. Because, on the modern welfare scheme, the rich do not decline in what is nowadays called *social capital*,[18] they are able to do much more than the poor with whatever income they earn, no matter how much their surplus is then taxed. In that case, a *social inheritance tax* is called for, the exemplar of which today is affirmative action legislation. Focusing on academic posts in science, it is not sufficient for graduates of first-rate institutions to be hired by third-rate institutions and vice versa; the difference in social capital remains and, in the long term, one would expect the former group to regain much of their original advantage. Instead, redistribution must occur in terms of who is trained at those different-quality institutions in the first place, with an eye towards enabling graduates to enter the employment market with roughly equivalent amounts of social capital and hence to be evaluated solely on their own merits. Here it is worth recalling that a person's 'social capital' is marked mainly by whether their background and training enable them to connect with others so as to realise their own goals: that is, the same knowledge content may contain different amounts of social capital, depending on the situations in which knowledge is able to bring about action.

Second Republican Strategy: Removing Cumulative Advantage from the History of Science

I have just proposed a general strategy for enabling individuals to overcome the principle of cumulative advantage. However, alongside that strategy must be a demystification of the very idea that, in the long term, some tradition or line of inquiry is bound to dominate, simply by virtue of the advantages it will have accumulated over its rivals. In the case of science, this idea appears most clearly in historically based doctrines of progress. No matter how postmodern we claim to be, the idea that science progresses by retaining the wheat and shedding the chaff of the past remains an article of faith in the public sphere. But in keeping with republican scruples, our strategy here will be to capitalise on historically unrealised possibilities to rework the faith. Specifically, I propose to reinvent the distinction between the contexts of *discovery* and *justification* of scientific knowledge. For people trained in history, philosophy and sociology of science, this call will undoubtedly bring back unpleasant memories of logical positivism, with its strictures against letting the idiosyncratic psychosocial origins of a knowledge claim interfere with the publicly inspectable means by which its validity is determined. The former—the realm of discovery—was by implication irreducibly irrational, and the latter—the realm of justification—steadfastly methodological. The distinction can be traced to William Whewell, the early Victorian polymath who coined the word 'scientist' as the name of a profession whose practice called for specific academic credentials.

In the 1830s, Whewell realised that virtually every major scientific innovation with industrial applications over the previous half-century had been made by people without university training. Whewell was concerned that if the humanistically oriented universities did not incorporate experimental inquiries into their curricula, they would be eventually forced to yield their authority to the emerging vocational schools and polytechnics that typically trained these innovators. Consequently, Whewell argued that scientific innovations were not fully understood unless they could be explained as the natural outgrowth of an evolving body of theoretically grounded knowledge. The ability to provide such explanations was the mark of a 'scientist', as opposed to the amateur inquirer. Thus, the medieval institution of the university received a new lease on life as the provider of justifications that went beyond the simple fact that a given innovation yielded immediate practical benefits.

In the 1930s, the discovery/justification distinction exchanged these elitist origins for more populist ones, with the invention of the 'genetic fallacy' as philosophy's first line of defence against the Nazis.[19] When the Nazis argued that the ethnic origins of a scientific innovator contributed to an innovation's validity, the logicians struck back that valid knowledge claims can come from any number of sources, thereby rendering an innovation's origins irrelevant to its ultimate validity. However, with the passing of the Nazis, the urge to stamp out the genetic fallacy has weakened. Indeed, the discovery/justification distinction is said to have been definitively deconstructed in the 1960s and 1970s by the 'historicist' strain in the philosophy of science, as represented by Thomas Kuhn, Imre Lakatos and Larry Laudan. According to these historicists, a research tradition justifies its continuation by the number of robust discoveries that are made under its auspices. On this view, a research tradition enjoys intellectual property rights over the knowledge claims it originates. Thus, if a scientist working in, say, the Newtonian or Darwinian research tradition happens to make an important finding, then the finding counts as a reason for promoting the tradition, and soon the impression is given—especially in textbooks—that the finding could have been be made *only* by someone working in that tradition. In other words, priority quickly becomes grounds for necessity.

Clearly the historicists presuppose a highly competitive model of scientific inquiry that gravitates towards the dominance of a single paradigm in any given field. They rarely countenance cases in which knowledge claims originating in one research tradition have been adapted to the needs and aims of others. One important reason is that ultimately the historicists believe that alternative research traditions are little more than ways of dividing up the labour in pursuit of some common goals of inquiry, such as explanatory truth or predictive reliability. Thus, they presume that there is some automatic sense in which a discovery made by one tradition is 'always already' the property of all—though access to this supposedly common property requires that one changes allegiances first. Consider the treatment of Darwinian evolution and Creation science as mutually exclusive options in the US public school curriculum. Although two-thirds of Americans who believe in evolution *also* believe that it reflects divine intelligence, such compatibility has yet to be seen as a philosophically respectable option, and consequently has no legal import.[20] But what exactly would be wrong with teachers trying to render biological findings compatible with the Creationist commitments of most of their students? One common answer is that the presupposition of a divine intelligence or teleology has retarded biological inquiry in the past and has not contributed to evolutionary theory since the time of Darwin's original formulation. Yet, the contrary presuppositions of mechanistic reduction and random genetic variation have equally led to error.[21]

In this context, republicanism arises as a pedagogical issue: Should students be forced to accept the current scientific canon as a paradigm (in the Kuhnian sense of a total ideology) that would deny the legitimacy of whatever larger belief-systems they bring to the classroom? Or, should students learn how to integrate science into their belief systems, recognising points of compatibility, contradiction, and possible directions for personal intellectual growth? If we favour the latter 'citizen science' perspective of republicanism over the former 'professional science' perspective of communitarianism, then we need to reinvent to discovery/justification distinction.

According to the old distinction, an ideally justified discovery would show how anyone with the same background knowledge and evidence would have made the same discovery. The role of justification was thus to focus and even homogenise the scientific enterprise through a common 'logic of scientific inference'. In practice, however, 'the same background knowledge and evidence' understated what was actually needed for people to draw the same conclusions, namely, involvement in a particular research tradition. In contrast, the new distinction I propose conceptualises scientific justification as removing the idiosyncratic character of scientific discovery in a deeper sense than the old distinction pursued—not simply the fact that a discovery was first reached by a given individual in a given lab, but the fact that it was reached by a particular research tradition in a given culture. In other words, the goal of scientific justification would be to remove whatever advantage a particular research tradition or culture has gained by having made the discovery first.

This project would safeguard scientific inquiry from devolving into a form of expertise, whereby, say, one would have to be a card-carrying Darwinian before having anything credible to say about biology. It would also have the opposite effect to the old distinction, in that it would aim to render a discovery compatible with as many different background assumptions as possible, so as to empower as many different sorts of people. Models for this activity can be found in both the natural and the social sciences. In the natural sciences, there are 'closed theories' (e.g. Newtonian mechanics) and 'dead sciences' (e.g. chemistry), which can be learned as self-contained technologies without the learner first having to commit to a particular metaphysical, axiological or perhaps even disciplinary orientation. In the social sciences, conceptual and technical innovations originating in one tradition have typically been picked up and refashioned by other traditions, so as to convey a sense of history as multiple parallel trajectories. It is just this cross-fertilisation that gives the social sciences the appearance of a field of unresolvable ideological differences—which, from a republican standpoint, is a *good* not a bad thing.

But perhaps the historically most interesting models of this renovated sense of discovery/justification at work are the hybrid forms of inquiry that have emerged as a defensive response to Western colonial expansion. In my book *Science*,[22] I consider the cases of modern Islam and Japan. In both, the instrumental power of the natural sciences has been neither denied nor anathematised, but rather systematically reinterpreted so that these sciences become a medium for realising the normative potentials of their respective cultures.[23] Along the way, some telling critiques of the historicist perspective on science are made. Basically, Islam criticises the West for not anticipating the destructive and despiritualising consequences of its 'science for its own sake' mentality, whereas Japan lodges the reverse charge that the West superstitiously clings to the stages undergone by its own history as a global blueprint for the advancement of science.

Whewell famously compared the relationship of scientific discoveries to their justification with tributaries leading to a major river. Sticking with the fluvial metaphor, I counter with the image of a major river opening up into a delta in which multiple traditions can make use of a body of knowledge that originated in only one of them. In Table 1,[24] I compare the implications of the two metaphors. In the next and final section, I follow Whewell's lead in regarding the university as the institutional site for the discovery/justification distinction to be drawn. However, given the distinction's republican renovation, the values invested in the university will turn out to be somewhat different as well.

Table 1. **Redrawing the contexts of the discovery/justification distinction**

	Old communitarian science	New republican science
Metaphor guiding the distinction	Convergence: tributaries flowing into a major river	Divergence: a major river flowing into a delta
Prima facie status of discovery	Disadvantage (because of unexpected origins)	Advantage (because of expected origins)
Ultimate role of justification	Concentrate knowledge through logical assimilation	Distribute knowledge through local accommodation
Background assumption	Discoveries challenge paradigms unless they are assimilated to it	Discoveries reinforce paradigms unless they are accommodated to local settings
Point of the distinction	Turn knowledge into power (magnify cumulative advantage)	Divest knowledge of power (diminish cumulative advantage)
Definition of contemporary science	Present is continuous with the future—the past is dead and best left to historians	Present is continuous with the past—the future is open to the retrieval of lost options

The University's Search for a Republican Identity: Enlightenment Lost

When the first universities were established in medieval Europe, academics dedicated themselves exclusively to the cultivation of the intellectual virtues, which included deciding who was mentally fit to undertake a disciplined course of study. But they did not minister to lay people's souls. They were *monks*, not *priests*. Priests, of course, were invested with just such a pastoral mission, which immediately expanded their sphere of accountability. They had to confront not only the Creator and their own colleagues, but also a lay constituency. This distinction between the monastic and priestly roles of academics, while easily ignored in the early days of the national university systems, has become more pronounced as the 'flock' has come to be dominated by some rather large wolves in sheep's clothing, especially funding agencies supported by either the taxpayer or private business concerns. Under the circumstances, it is not surprising that a wedge has increasingly been driven between the alleged universality of the knowledge produced in such settings and the multiply interested knowledge producers themselves. In other words, the tensions underlying civic ideals and public goods that usually destroyed past republican regimes have been regularly enacted in the history of the university.

These tensions can already be found in the original period of Enlightenment, the 18th century, which resembled the Protestant Reformation in its attempt to retrieve the critical functions of inquiry from institutionalised captivity to its clientele. For the most part, the champions of Enlightenment were freelance writers who regarded the universities as little more than propaganda ministries for *les anciens régimes*. They strove to free 'science' from its alleged 'scientists'. The university itself enabled such a clear distinction between 'product' and 'producer' precisely because, as a state function, knowledge was provided with a public character that opened it to scrutiny by those outside the immediate sites of knowledge production. In short, what enabled knowledge to serve some interests outside the university's precincts also enabled it to be criticised by others equally outside its precincts. It is a tribute to the Machiavellian genius of Wilhelm von Humboldt and the German architects of the 19th century 'research university' that the Enlightenment was so effectively co-opted into the strictures of academic life that by the beginning of the 20th century the critical impulse that had animated Voltaire and Marx had become safely sublimated in the cramped prose of Max Weber. Of course, the heirs to the Enlightenment spirit had actively conspired in their own co-option by succumbing to the siren song of positivism. It is worth recalling these inconvenient facts, lest we be seduced by an airbrushed history of the university, one that suggests that the moment of Original Sin occurred only when academics tasted the forbidden fruit of the atomic bomb in 1945.

Moreover, we must do more than simply recall the lost history. We need to reinvent for our own times those aspects of the original 18th-century Enlightenment that have never found a place in the canonical histories of the university. What I mainly have in mind here is the belief shared by most correspondents in the Enlightenment's 'republic of letters' that the success of Newtonian mechanics as an empirically demonstrable form of knowledge had brought the major period of esoteric inquiry to a close, and that intellectuals had best spend their time ensuring that the 'new science' did not further consolidate into a secular priesthood more virulent than the one traditionally ensconced in the universities. In other words, the Enlightenment wits would have rejected wholesale Kuhn's popular puzzle-solving picture of 'normal science', whereby once anchored to a paradigm, scientists pursue an increasingly autonomous and narrowed range of inquiry. Some of the wits would have been attracted to the German finalisationist school of critical science policy, briefly in vogue when Jürgen Habermas directed the Max Planck Institute in Starnberg in the 1970s.[25] Here the state steers the course of science in more practically relevant directions once a critical mass of knowledge has been reached.

However, perhaps the truest vision of Enlightenment science was the ideal epitomised by such American founding fathers as Benjamin Franklin and Thomas Jefferson, who practised science as part of living a cultivated life, just as one might also write poetry and play music. In this way, science is not merely downsized but integrated into each individual, thereby dissipating the mystique of 'divinity' that many British clerics and eulogists attached to Newton's original accomplishment. This truly democratised science would thus be *science secularised*.[26] There have been periodic stirrings of this spirit in the nearly two centuries since the research university embalmed the Enlightenment. However, a sense of the speed with which this embalming occurred in the early 19th century may be had from the dubious place that Johann Wolfgang von Goethe occupies in the history of science today, despite having practised science in exactly the spirit cultivated by the likes of Franklin and Jefferson. Though now known primarily as The Great German Poet, Goethe spent most of his life as a civil servant; and, were it not for the academic co-option of critical inquiry in the 19th century, he might well have been remembered mainly as a distinguished intellectual descendant of that other great German civil servant, Gottfried Wilhelm von Leibnitz. In the 20th century, Ernst Mach, Otto Neurath, Karl Popper and Paul Feyerabend come to mind as representatives of this repressed tradition. For a sense of just how repressed this side of the Enlightenment has been, consider that notwithstanding postmodern portrayals of the Enlightenment as the epitome of academic respectability, most of the 18th-century wits flouted scholastic writing conventions in open defiance of academic claims to epistemic privilege. Indeed, their attitudes towards inquiry exhibited much of the playfulness that postmodernists today claim for themselves.[27]

At this point, let me caution against one interpretation of the Enlightenment legacy, namely, that the university should be in the business of mediating society's cultural clashes, *but* without taking a stand of its own. Allowed to take its course, this avowedly neutral form of mediation would effectively put the university in the business of increasing the complexity of an already complex world by inserting yet more layers of (presumably discursive) mediation. To my mind this would undermine the university's republican potential. As the flagship institution of societal reproduction, the university is entrusted with distilling the various cross-currents in a society's collective life into a 'heritage' that is worth imparting to its next generation. The very idea of a 'canon' may have become the object of much abuse in these postmodern times, but it stands metonymically for what the university is all about. Of course, the contents of a given canon (in, say, literature or philosophy) have tended to outlive their usefulness. However, *pace* postmodernists, the solution is not to eliminate the canon and replace it with a plethora of cross-cutting discourses that either match or increase the level of societal babble. Rather, the solution is to revise the canon regularly so as to highlight aspects of social life that students might otherwise ignore or devalue in the normal course of their lives. In other words, *the university should simply institutionalise oppositional consciousness.* It should become the clearing-house for all the voices that would otherwise be silent or muted beyond recognition, with the understanding that these will change as the power relations in society change. In that sense, 'affirmative action' would be built into the institutional design of the university, rather than simply be an unwanted constraint on its 'natural' activities (i.e. reinforcing the dominant traditions). Only in this way can the university continue to play a necessary role in the processes of societal reproduction—namely, by reproducing dissent and difference—while retaining its independence. However, there is a price to pay for the university agreeing to this new social contract, which will become evident below.

An important consideration in my thinking is that I believe that most of the alleged 'complexity' claimed for the postmodern condition has been manufactured by academics themselves, often unwittingly and in response to circumstances not entirely under their direct control. When I say that complexity is 'manufactured', I mean that it does not refer to something deep about the nature of a reality that is, in principle, independent of the collective activity of human beings. Thus, while we may have been caught off-guard by the realisation that we now live in an especially complex world, nevertheless the processes of 'intermediation' involved in constructing this complexity can, and should, be reversed. Below I enumerate four sorts of intermediation that together constitute the pervasive sense of complexity that haunts much of contemporary social theory. Reversing these forms of intermediation will force defenders of the university to take some hard decisions, yet ones that are necessary for the return of the repressed side of the Enlightenment, the side that would bring us back to civic ideals and public goods.

I shall deal with the first two forms of intermediation in relatively short compass, since I have written about their historical significance elsewhere.[28] First, there is *economic intermediation*, which is encapsulated in the saying 'High entry costs make for high exit costs'. Once enough material and human capital have been invested in a line of inquiry, it becomes difficult to justify its discontinuation, especially if it reaps reasonable benefits for those pursuing it. Moreover, this continued investment lowers the chances that others will be in a position to contribute to the inquiry unless they enter through the already established channels. Here the sheer concentration of resources can easily provide the illusion that the field exhibits a level of complexity that can be fathomed only by the keenest and most dedicated minds. In fact, this complexity is often traceable to the ingenuity needed to translate legitimate areas of cognitive interest into the confines of obscenely expensive machinery and excessively narrow technical skills.

Second, there is *functional intermediation*, which results from the implicit social contract that science has made with the state in the 20th century, whereby the research frontier is allowed to advance apace, on the condition that scientists also contribute to the normative reproduction of society. However, the trajectories of these two functions have increasingly diverged over the course of this century. Perhaps the clearest case in point is the continued use of thinly veiled intelligence tests as a social sorting mechanism, despite their methodologically suspect character. This dissonance in the knowledge system has opened a space for all manner of people to broker the difference through activities ranging from the writing of pop psychology to the conduct of so-called policy-relevant research.

Third is *temporal intermediation*. Sometimes a 'Before & After' sense of history is invoked to highlight the complexity of the current knowledge system. But as a matter of fact, the difference between the time the experimental natural sciences were regarded as sufficiently theory-driven to be deemed fit subjects for the university curriculum and the time they started to take the lead in forging links between the university, industry and the state in large-scale research projects is no more than *one generation*. In the case of the vanguard nation, Germany, this period spanned the third and fourth quarters of the 19th century. The start of this period was marked by the first nationally adopted textbooks in experimental physics, Kuhn's ultimate criterion of normal science; the end by the establishment of the first hybrid research establishments, the Kaiser Wilhelm Institutes. In the latest lingo, 'Mode 1' and 'Mode 2' knowledge production would, then, seem to have been virtually joined at the hip at birth, not separated by several centuries and incommensurable worldviews. However, to hear the 'Modists' tell the history, Mode 1 and Mode 2 have quite distinct origins, the one in the remote past and the other in the recent period. Kuhnian paradigms had been in place at least since the founding of the first autonomous scientific societies in the 17th century (and maybe were present even in the ancient and medieval worlds), while the social hybridisation of the scientific research agenda did not take root until the atomic bomb project (and really did not precipitate a crisis in knowledge production until the post-Cold War period).[29]

What is going on here? In essence, the Modists present a *stereoscopic* view of the history of knowledge production that creates an illusion of temporal depth by driving a wedge between two rather closely connected developments, and then reading the earlier one in terms of what might be plausibly said to have led up to it, and the later one in terms of what might be plausibly said to have followed from it. The result is the appearance of deep rupture where there is none. Ironically, the illusion is maintained because, despite talking a 'hard-headed' policy-relevant line, the Modists systematically ignore salient features of the institutionalisation of inquiry. To take just one example, it is misleading to classify a chartered scientific society and a publicly maintained university as 'autonomous' research institutions in the sense of Mode 1, if only because they are not accountable for their activities in the same way. More importantly, the Modists' stereoscopic historical perspective leads them to overlook that it became easy for academics to move into Mode 2 knowledge production processes once they identified the source of their 'autonomy' with their discipline rather than the university as such. Laboratory scientists were not included in either the medieval or the Humboldtian conception of the university as the natural home for inquiry. The corporate consciousness of scientists was rather modelled on such emergent 19th-century formations as professional associations and labour unions. For laboratory scientists, there has always been the sense that the university is a more-or-less hospitable space for doing business but not constitutive of their identity as autonomous inquirers. Consequently, scientists who shifted from Mode 1 to Mode 2 have generally done much less soul-searching than might have been expected, given the alleged chasm that separates the two modes.

The last but by no means least form of intermediation is what may be called *collegial intermediation*. These days it is not sufficient to know about the world for oneself; one must first know about others who know about it. In crude terms, it is the problem of having to read what everyone before you has written before you are permitted to say anything for yourself. Much of the historical uniqueness that is claimed for the 'postmodern condition' is probably traceable to this easily taken-for-granted feature of academic life. It involves treating the expansion of the universities since the end of the Second World War as if it were in direct response to some profound change in our understanding of reality, if not the nature of reality itself. However, objectively speaking, the world is probably not changing more than ever before. Rather, there are more academics looking at it, and because they are rewarded for their 'distinctiveness', as Pierre Bourdieu would put it, they have made careers by adding new 'perspectives', probably beyond necessity. I am not even convinced that, all things considered, the world has become a more uncertain place in which to live—a common subjective indicator of contemporary complexity. What is incontrovertible is that the sheer availability of alternative perspectives has made it difficult to bring closure on matters that require a specific course of action. Not surprisingly, then, social movements that have rallied around the postmodern exposure of 'global uncertainty' are more effective in blocking the actions of others than in taking action for themselves.

In sum, I propose that the university recover its Enlightenment promise by becoming, so to speak, a medium of *disintermediation*. It would thus aim to reduce the complexity of the social world as part of an overall strategy of empowering citizens to pursue common ends. In terms of the four forms of mediation mentioned above, it would mean that the university would reduce (1) its financial dependency, which implies discouraging research that requires enormous expenditures; (2) its role in normalising the social order, which implies divesting many of its credentialising functions; (3) its sense of historical rupture, which implies determining its own fate *vis-à-vis* so-called 'irreversible' forces of change; (4) its sheer profligacy of positions, which implies that academics take a more strategic attitude towards the larger society (rather than simply each other) to become a collective agent of oppositional consciousness. The resulting institution would be, in many ways, more ascetically maintained than our current 'multiversities'. But with that asceticism would come an ability to speak *to* power without having to speak *for* it. The university would at last become a republican institution.

The Future of Futures Studies as a Function of the Future of Science

As a relative newcomer to futures studies from history, philosophy and social studies of science, the aspect of the field that I find most surprising—and not a little ironic—is its relative lack of attention to the sense of historicity that invariably accompanies any judgement of what the future may hold. How must one be situated in history in order for the future to appear as it does? We have just seen that the future identity of the university rests crucially on how its past is projected into the future. The point was also explicitly made when I considered the difference in the past's relationship to the present as advanced by the communitarian vision of science associated with Kuhn and by my own republican science alternative. For his part, Kuhn notoriously described science's sense of its own history as 'Orwellian',[30] which means that science's present is presumed to be the culmination of its past and that the future will be a logical extension of the present. Of course, this streamlined presentation of the history of science massages, if not outright erases, dissident traditions whose divergent tellings of the past would cast a simple extrapolation of current trends in a much less favourable light. Such a mutual determination of the past and the future is captured by Raymond Aron's elegant phrase 'The past is never definitively fixed except when it has no future'. I have tried to work out the complementary principle with regard to science: *The future is never definitively fixed except when it has no past.*

The distinctly focused and progressive character of the history of science that has inspired so much futures studies research is based at least as much—if not more—on the remarkable degree of uniformity among scientists as to what should be preserved and discarded from their past as on the genuine foresight of heroic visionaries and utopian planners.[31] Indeed, I would argue that the sociological attribute most closely connected to this linear sense of history is not any clear sequence of scientific advances but simply the efficiency with which structural amnesia is imposed on the collective memory of science. In a book that I am currently completing on the origins and impacts of Kuhn's *The Structure of Scientific Revolutions*, I observe that the Orwellian sense of history that Kuhn ascribed to scientists was partly inspired by Alfred North Whitehead's famous quote 'A science that hesitates to forget its founders is lost.'[32] In closing, let me suggest a relevant counter-quote from another Harvard sage of the early 20th century, George Santayana, who declared, 'Those who cannot remember the past are condemned to repeat it.'[33] After Kuhn, it has become common to think that history needs to be left behind if science is to make progress; but it may be that what Kuhn and his followers have taken to be the mark of intellectual resolve on the part of the scientific community is really, as Santayana thought, nothing more than a sign of either infancy or, more likely, senility. Ironic though it be, a forthright attempt to make the past contemporaneous with the present may be the best strategy for progressive thinkers in any field of inquiry to keep the future forever open.

Notes

1. Francis Fukuyama, *The End of History and the Last Man*, New York: Free Press, 1992.
2. T. S. Kuhn, *The Structure of Scientific Revolutions*, 2nd edn., Chicago, IL: University of Chicago Press, 1970.
3. Nico Stehr, *Knowledge Societies*, London: Sage, 1994.
4. Stephen Cole, *Making Science*, Cambridge MA: Harvard University Press, 1992.

5. Steve Fuller, *Science*, Milton Keynes: Open University Press, 1997; Steve Fuller, *Philosophy, Rhetoric and the End of Knowledge*, Madison: University of Wisconsin Press, 1993.
6. Philip Pettit, *Republicanism*, Oxford: Oxford University Press, 1997.
7. James Bohman, *Public Deliberation*, Cambridge, MA: MIT Press, 1996; James Fishkin, *Deliberative Democracy*, New Haven, CT: Yale University Press, 1991.
8. Jon Elster, 'Constitutional Bootstrapping in Philadelphia and Paris', *Cardozo Law Review*, 14: 549–75, 1993.
9. Fuller, *Science*, ch. 4.
10. Roberto Unger, *What Should Legal Analysis Become?*, London: Verso, 1996.
11. Karl Polanyi, *The Great Transformation*, Boston, MA: Beacon Press, 1944.
12. J. G. A. Pocock, *Virtue, Commerce and History*, Cambridge: Cambridge University Press, 1985.
13. J. G. A. Pocock, *The Machiavellian Moment*, Princeton, NJ: Princeton University Press, 1975.
14. Robert Merton, *The Sociology of Science*, Chicago, IL: University of Chicago Press, 1973.
15. Karl Popper, 'Normal Science and Its Dangers', in I. Lakatos and A. Musgrave (eds.), *Criticism and the Growth of Knowledge*, Cambridge: Cambridge University Press, 1970, pp. 51–8.
16. Kuhn, *The Structure of Scientific Revolutions*.
17. Merton, *The Sociology of Science*.
18. James Coleman, *The Foundations of Social Theory*, Cambridge, MA: Harvard University Press, 1990.
19. Morris Cohen and Ernest Nagel, *An Introduction to Logic and the Scientific Method*, London: Routledge & Kegan Paul, 1934.
20. Stephen Carter, *The Culture of Unbelief*, New York: Doubleday, 1993.
21. John Horgan, *The End of Science*, Reading MA: Addison Wesley, 1996, pp. 114–42.
22. Fuller, *Science*, ch. 6.
23. Ziauddin Sardar, *Explorations in Islamic Science*, London: Mansell, 1989.
24. Steve Fuller, *Being There with Thomas Kuhn: A Philosophical History for Our Times*, Chicago, IL: University of Chicago Press, 1999, ch. 1.
25. Wolf Schaefer (ed.), *Finalization in Science*, Dordrecht: Reidel, 1984.
26. Steve Fuller, 'The Secularization of Science and a New Deal for Science Policy,' *Futures*, 29: 483–504, 1997.
27. Anne Goldgar, *Impolite Learning: Conduct and Community in the Republic of Letters, 1680–1750*, New Haven, CT: Yale University Press, 1995.
28. Steve Fuller, 'Towards a Philosophy of Science Accounting: A Critical Rendering of Instrumental Rationality', *Science in Context*, 7: 591–621, 1994.
29. Michael Gibbons *et al.*, *The New Production of Knowledge*, London: Sage, 1994.
30. Kuhn, *The Structure of Scientific Revolutions*, p. 167.
31. Fuller, *Science*, ch. 5.
32. Fuller, *Being There with Thomas Kuhn*, ch. 6.
33. George Santayana, *The Life of Reason*, vol. 1: *Reason and Commonsense*, New York: Scribners, 1905, p. 284.

14
Futures Studies and the Future of Art

SEAN CUBITT

Of all the issues that face us as we look towards the millennium, perhaps the least pressing is the question of art. Yet art, existing in every society across the world, can help us understand the call Ziauddin Sardar makes for a radical rethinking of technological, global and scholarly futures. After all, it was only a small thing, the observation of the moons of Jupiter, a matter of no consequence in the great ways of the world, that opened the doors for modern science, and brought about the downfall of the Christian West. From such minutiae, great changes spring: this is the power of specialisation, which has powered the science and technology of five centuries, and even now leads the way in the development of a controlled and administered future.

In particular, I want to talk about the media arts, a smaller segment yet, but of great importance to all our futures. Again, Professor Sardar gives me a start, with his questioning of the ways in which the media transnationals have begun the process of globalisation simultaneously with their colonisation of the future. The two go hand in hand: globalisation is not thinkable without a business strategy based on the futurological principle of return on investment. But if you wish to invest in communications, you must be swift: today's news is a valuable commodity; yesterday's news is just a library. And at a certain point, the idea of return on investment gives way to the idea of protecting an investment. It is at this point that the worst aspect of future planning takes over, for all too much of the administration of the future, the management of change, is directed towards maintaining the status quo, ensuring that nothing is transformed between now and then.

This is in part a matter of content; of ensuring that CNN is shown everywhere in the same way with the same stories and the same perspectives on them. But it is also, and very importantly, a matter of standardising the hardware. The economics of scale demand that we all have television screens of the same shape. We are not encouraged to use projected images instead of TV sets, so we could watch TV together rather than as a domestic technology. Many of the new machines—the walkman and the computer are two of them—are designed for the exclusive use of a single person, reinforcing the Western ideologies of individualist competition at work, and individualist gratification in leisure. The computer screen echoes the TV screen, and the stereo headphones echo the lonely silence of the cinema viewer lost in the darkness.

The computer adds more difficulties. The QWERTY typewriter keyboard was never a good design: in fact, it was engineered to slow typists down, so that the keys of the early machines would not get tangled up with each other. But the design became the norm in the office-equipment revolution of the early 20th century that brought women into the workplace, displacing the old hand-written craft of male employees and cheapening the labour of writing, while also standardising not just the typeface, but the process of producing it. Standardisation made all women typists interchangeable, components in a standardised office bureaucracy, and able to move from firm to firm without retraining. The difficulty of the keyboard layout, the physical skills it demands (making it unsuitable for anyone with minor disabilities, arthritis or simply ageing hands) and its universalisation even in areas where the languages are simply unsuitable for this kind of alphabetic inputting are clear examples of colonialism at work not only in the content but in the design of our media technologies.

Perhaps the other single greatest achievement of the new media technologies is their success in producing English as the international *lingua franca* of technology, engineering, software, computer manuals and the Internet. Though I confidently expect that spoken English, already a complex family of often mutually incomprehensible dialects, will evolve into a series of new languages, written English will almost equally certainly evolve, if at all, only very slowly in comparison. As a result of its standardisation in the interests of comparable skills, corporate communication and the maintenance of a cultural hegemony even when the economic ascendancy of the West is in decline, English will become the Latin of the next century.

It is in this context that I want to talk about art, the media arts especially. There are many issues to be raised in addressing the future of the media: issues of ownership and control, censorship and accountability, democracy and access. But there is also the question of what the media are made of, and by looking into some work that artist–engineers have undertaken, I believe we can catch a glimpse of the kind of experimentation that may yet produce the possibility for a new and open future. But first, we need to take a little detour into history.

A Historical Detour

What is art, this small thing, this margin of the great doings of the world? For some, it is consolation, the clean, clear world of an enfolding glow that puts to rights, or at least gives comfort, amid the wreckage of everyday life. There exists a huge market for these comforts and distractions, for music that enacts the wistful nostalgia for youth and beauty, films that yearn for high emotions we have no time for in real life, paintings of the heroism that we never get to live out. But what are these cultural products today but the ordinary commerce of popular cultures, the endless recycling of operas and symphonies, old master paintings and classic novels turned into TV and films, consumed in the same way as we consume our food: a pornography of the soul. An art of consolation is a blanket you wrap yourself up in so as not to feel the chill winds of reality.

Already in the 18th century, Kant declared the gap between beauty and interest: that which is truly beautiful is without 'interest' in the sense that it does not offer to solve or resolve the problems and contradictions of life, to satisfy desires or answer questions. Kant invented modernist art, the art of the pure object, an art freed of the necessity to serve God or King, free to pursue its own life and its own logic. It was this new understanding that brought about the possibility of abstract art in the West (although, of course, it would arrive there, on the back of 19th-century Orientalism, hundreds of years after the development of complex abstract visual languages in cultures from Machu Picchu to Great Zimbabwe). This new art would claim to be the very shape of freedom, the great refusal of whatever exists today. Modernist art would be the negation of everything that exists.

Even though art has been taken over by commerce and national institutions, there are activities that have a special status, that we call 'art' not because they are in galleries and have frames around them, but because they inhabit a different space to the everyday world. Art acts in many ways like the devotional sculptures, paintings and architecture of more spiritual times and cultures: to create a sense of a world beside, outside, beyond—a beyond that, for a secular age, is always future.

This is not to say that art depicts the future. Even when it shows us death, the only future we can be assured of, or the afterlife, it is not what is shown that is important but the manner of its showing. The best art, I would say, is the art that constantly evolves out of what has already become acceptable and comfortable, already part of that everyday that it is art's business to reject. We no longer find Picasso shocking—on the contrary. Art is the process of transformation itself, of transforming the raw materials of life into art, and of transforming the raw materials of art into a newer art, a different and always displaced work. Art is not about objects to be purchased or hung in national mausoleums. It is the living practice of transforming the everyday into its negation.

Of course, one can be cynical about this too. In certain instances, art is just the research and development wing of the media and communications industries. Every new technique that artists can come up with will be turned sooner or later—usually sooner—into advertising billboards, film posters, television programmes and comic books. But this is a great secret: that no technique is intrinsically good, intrinsically radical, intrinsically anything. All techniques, like all technologies, can be turned to good or ill, the purposes of humanity or the purposes of corporate strategy.

We are pretty sure that we know what the media are. Film, radio, television, records, computers: we have had a century to get used to the familiar devices of the modern, mechanical culture. Yet the oldest of the media are scarcely 150 years past their first appearance; video is a scant 30 years old; and popular computing and network communications barely have a decade under their belts. At the same time, in each of these forms, the bedrock of the medium's development has been settled within fifteen or twenty years of its first appearance, economically, institutionally, politically and technically. The first recognisable Hollywood film arrived promptly twenty years after the invention of the medium, by which time the studio system, the star system, the rectangular screen, the 35mm standard size of film strip and the practised division of craft labour were already ensconced. Yet it was none of these that made Hollywood the global dream factory. It was the curious history of cinema exhibition that made it central, for every one of the major studios evolved from the distribution networks that controlled the flow of films around the huge internal market of the USA, and allowed them to colonise all the Americas within a decade.

Dadasaheb Phalke was no less an artist than D. W. Griffith or Charlie Chaplin. What he lacked, however, in the conditions of the British Raj, was a distribution network. Instead, we hear of him bringing home bullock carts of anna coins himself from screenings he and his family arranged. Similar stories occur throughout the century-long struggle for a 'second' cinema of bourgeois art, and a 'third' cinema of popular political film. Even the international avant-garde, otherwise so adept at circulating their work, were reliant on the galleries and museums for occasional screenings. The lessons of Hollywood were learned well in the USA, as indeed they were, rather differently, in Europe and across the world, when it came to establishing broadcast networks. In the USA the electrical manufacturers, and in most other countries the state, established effective monopolies or cartels to administer transmission standards, admission to the industry and the types of programme content. Even today, it is hard to think of more than a handful of television events that one could call 'art', certainly in the sense of a disinterested aesthetic. The relative cheapness of record production has allowed a far greater growth in marginal and aesthetic cultural production, but never to the detriment of the transnational corporations.

Today, as these corporations are engaged in the convergence of media and companies, the launch of a major product occurs simultaneously across film, TV, records, computer games, clothing, advertising, fast food and theme parks. Steven Spielberg's 'The Lost World', already the biggest money-maker in cinema history, is an example of this cross-fertilisation of markets. When we now speak of multimedia, we mean both this ability to launch a product like Batman in every available medium, and also the specific convergence of media in digital formats like CD-ROM and the Internet. This convergence is almost entirely dominated by a handful of corporations, and in many respects, the Western liberal academy has fallen in line, with its detached, ironic and ultimately defeatist stance of 'postmodernism'.

Manuel Castells, in perhaps the most important work to be published yet on the networked society,[1] suggests that Africa is surplus to the requirements of the globalisation process. An information economy has far less need of raw materials, education levels are below the minimum needed for the information and technology industries, poverty makes it an uninteresting market to exploit for the typical products of the new age, and the only thing that keeps the aid coming is the fear of metropolitan countries that they might have to feed and house vast numbers of migrants heading towards the wealthy economies. This numbing prospect suggests not only the failures of the postmodern academy's dematerialisation, but the scale of the problems faced by any attempt to reconsider what the new media must achieve.

Of course, no revolution in the media, no matter how wide-ranging, can have more than an indirect impact on economic and political problems of continental scale. Yet, just as the consolidation of globalised corporations has impacted on the development of a global information economy, exclusive of Africa, so the development of new media forms may make its contribution to the building of a decent future. At its core will be the justice of the distribution systems. In some ways, the struggle between modernism and postmodernism has been the struggle between an interest in production and an emphasis on consumption, with modernists arguing for control over the means of production and postmodernists arguing that the mass audience is interested only in a perhaps vaguely creative form of consumerism. But what seems to be changing, potentially at least, is the possibility of centring attention on the mode of distribution, most of all because, with Internet, the means of production and distribution are the same.

The history of publishing should make us pause before leaping into any utopianism here. Small presses have had a wonderfully diverse history, for good and ill, since the invention of the printing press, yet they too have always been kept away from their audiences by the business domination of news-stand and bookstore sales. Today, anyone with a connection and a little easily acquired skill with software packages can put together an interesting and entertaining web site. The problem is that the media corporations have also noticed the burgeoning market-place in cyberspace, and are moving in hand over fist, with bigger pictures, louder songs and all the targeting that market research can bring them. Vast conglomerates comprising four or five of the largest global corporations are lining up to take over their slice of the new frontier, with all the sense of generosity and grace that imperial adventurers have been renowned for over the last five hundred years. Combining telecommunications, hardware and software companies and information and entertainment providers, the new players in the global market are out to take control of, and to standardise, the running of the Net. It is ironic that the only political organisation for networked activists is concentrated on the USA's fetish for free speech, and sees its only enemy as the US government, while all the time the remaining area for soapbox orators is shrinking by the hour in the face of corporate expansionism.

There are a variety of strategies that one might propose to keep open the possibility of an open and developing Internet. People will struggle within corporate structures for the best and freest programming. Alternative networks will spring up and fall back, to be replaced by others or co-opted by the dominant cartel. Hackers, the Internet's guerrillas, will sabotage the big players. In short, there will be resistance from within and without the big companies.

But although resistance has become one of the buzzwords of 1990s cultural studies, so much so that it is enough to describe something as 'resistant' to ensure that it will be seen in a positive light, resistance alone is not enough. The business of building the future cannot be undertaken by placing yourself at every turn in relation to the dominant. In fact, to resist the dominant is a way of multiplying its sense of its own domination, for domination thrives on the recognition of its subordinates. The question facing the media arts in the age of cheap global electronic publishing concerns the possibility of acting Otherwise, of distributing Elsewhere, of alternative networks and communities, on the global scale now made possible by computer-mediated communications.

The media arts offer a sense of how such networks might evolve. A favourite example of mine is the music of the African diaspora. A beat that originates in Ghana is transformed in Cuba, given a ragga twist in Jamaica, travels to New York, gets sampled in Detroit, plays in a dance club in London, where it is given a soul-boy edge, and gets picked up again in Paris and travels back to Ghana utterly and completely transfigured to begin its journeys once again. At no point in the cycle is the tune anyone's property. At no point is it fixed into a rigid and unchanging form. No one controls the network, which works on every message as a raw material for reinterpretation. Examples could be multiplied wherever a shared faith, a common cultural tradition, the merest ghost of a territorial loyalty ties together the dispersed communities that create the existing parallel networks that connect the world, beyond the boundaries of corporate culture.

Such alternative networks are everywhere on the Internet, communities of a few dozens or hundreds sharing their interests and their hobbies, an infrastructure that has the unrealised potential to link people on a global scale. But there remains a further aesthetic point that debars this from achieving its potential: the apparently fixed nature of the one-to-one relation we have, not with our Internet interlocutors, but with the white box on the desk. The hardware of telecommunications and the computer itself are designed to serve the needs of the global bureaucracy, not the desires and interests of the bulk of the population. At events like the International Symposium on Electronic Arts (ISEA), Arts Electronica and Video Positive, artists are busy deconstructing those apparently immutable boxes and their equally immutable wired connections, creating absurd, funny, poignant, tragic interfaces to supersede the dead hand of the white box. Most of all, more and more of them are devoted to multiple users—to removing the centrality of the technology, and replacing it with relationships between people.

This is the most profound lesson of diasporan communications networks: that they are about people, not things. Not only does this allow artists to rebuild existing machines; they are also making their own devices, frequently out of the simplest materials—old carpets, security lamps, broken televisions. Still more, they are throwing away the ideology of efficiency in communications, demanding the right to misinterpret and misunderstand, to treat online dialogue not as bureaucratic command structure, but as a space of cultural translation where misunderstanding is as fruitful and creative as truth (a fine example is the Webstalker, an anarchic browser that troubles the boundaries between art and engineering, the amateur and the professional).[2]

To be human is to communicate. Our making and working, loving and dreaming, buying and selling are communicative acts. Some communications are richer than others—making love gives more of us than collecting a pay cheque. And today, our communication is more often mediated through things—possessions, financial transactions—than through the broad bandwidth expressions of face-to-face fighting or sex or discussion. All communication is mediated, but some of those mediations—bodily, vocal, visual contacts—are so profoundly implicate in the construction of communication as to render indistinguishable the persons involved. When mediation is consolidated in objects—letters, paintings, music—it adds to these mutually complicit communications a wealth of additional, formal compositional tools, but when such objects are, as they have been in modern Europe, rendered autonomous, they act not as complements to the delicious intimacy of confrontation, but as boundaries between speakers, and latterly, in an age of information, between senders and receivers. And certain central mediating objects—most of all cash—contain almost no content at all, but merely mediate, offering only to place people in antagonistic relation to one another, enacting a separation that they can only experience. The contemporary history of communication is the dialectic of communicative mutuality and communicative mediation.

Communication is inherent, but it is also necessary, and its necessity makes clear a further property of the communicative theory of humanity: its ecological standing. For no one stands alone. If I reach out to touch the world, the world touches me back. If I look out into the world, I inscribe myself into the field of the visible, offering myself to sight. My mind is not, despite the arguments of the now triumphant cognitivists, bound inside the bone box of the skull: it extends through the nervous system into the life-world that touches and illuminates me. What strikes me most about this phenomenological relation is my dependency on the world. It is no secret that diet, water and air, noise and light impact on psychological states. The human is unthinkable without the world. For the last hundred years or more, that world itself, for most of us, has been human: the artificial world of the city, of the media, of social and cultural histories and environments. That is the human world on which, in the first instance, we are dependent. Today, that world is global. The tendrils of communication have thickened into trunk routes: the collapse of the Thai bhat means fewer East Asian students, which means my university earns a million or so less pounds next year, which means I cannot increase student numbers, and the local economy doesn't benefit from them, and the local culture gains no depth of diversity. Communication takes place within, and indeed is commensurate with, the global human ecology.

Because of my dependency, I must take responsibility: as I am affected by the sum of others, they are affected by me, and I will be the recipient of the ripple effect of my contribution to the global human ecology sooner, given the effects of proximity, rather than later. This is a reason to be critically alert in communication (not, by any means, to retreat from communication, as though such a thing were possible), and to question all attributions that we make customarily about all our interchanges. For example, we cannot be cosy about art: the word, the concept, the institutions, the practices of art cannot be considered sacrosanct. Walter Benjamin, in the dark hours of 1934, when both Hitler and Stalin had declared the end of the modernist experiment, argued that 'a writer who does not teach other writers teaches nobody'.[3] We might usefully slip this towards a redefinition of artistic practice: the artist who does not enable creativity in others is no artist. This is the responsibility born of the art-maker's debt to all those whose arts have impacted on her. But if this is our definition, it raises the question of the limitation of art to a small range of activities and concerns. I hold up as exemplary Kai's Power Tools, a software device for producing whizzy digital effects in computer-manipulated images; the bicycle and the lending library, the two most elegant products of the 19th century, true enablers; and Frank Annett, the English teacher who showed a generation how feeling and understanding did not have to be divorced; but also the art practices of Duchamp, Hendrix, Stein and a thousand others who have added to the repertoire of ways to think and feel, and most of all to communicate.

My point is not to devalue art, but to revalue the broad range of cultural practice. Writing and reading, any good teacher will instil in their pupils, are not just skills to acquire so as to pass through school with the minimum pain, or to emerge suitably equipped to survive the cut and thrust of adult life, but a lifelong source of delight, with or without a professional interest. The initial problem is one of education. We do not teach the making of sound or pictures in ways commensurate with the technologies that dominate communication. The failure to teach young and old to write the media as they learn to read them has two roots. One is the reluctance to overload a school schedule already groaning with the complex skills demanded by contemporary life, or to demand of teachers that they acquire yet another array of underpaid and undervalued skills. The other is a commitment, extending throughout the art education establishment, to media more appropriate to the 18th century than the 21st. I speak with the conviction of the convert: for four years as National Organiser of the Society for Education in Film and Television, I helped promote, across the UK, education about the media, but not the practical skills of media-making. Today, it seems curious to devote the majority of art school resources to painting and sculpture rather than photochemical and electronic imaging. And sound is in worse straits: only music is offered in any form, and that within a rigidly historicist and ethnocentric mode: eighty years after Russolo, where is the Art of Noise? Where are the conservatoires (and how revealing that title is) that teach gamelan or raga? The conservatism of public sector education has effectively denied to the mass of the population the pleasures of writing in light and sound.

This absurd institutionalisation of illiteracy in the major media of the age has as correlative the professionalisation of communication, a professionalisation rendered more damaging by the tendency most of us have to imitate the rhetorical and intonational patterns of the professional media when called on to speak or act in public. A professional media is as dangerous as a professional polity: both demonstrate a willing collusion in the demolition of democracy, which, if I am right, is the political state of a society in which as many of its citizens as possible are communicating in as many channels as possible. Only then do we genuinely produce public opinion, not that pale simulation of the five-yearly poll and the questionnaire. (Of course, we must also train professionals, responsible professionals: we cannot afford to leave politicians and programme-makers utterly ignorant, or admit defeat inside the beast, but we must also recognise the beast for what it is.)

Politics as it exists today is a struggle over the meaning of the future, the mode of futurity, between those who wish to control, to own the future, and those who seek its freedom, and in its freedom, their own. The future, as Levinas says, is other: 'the very relationship with the other is the relationship with the future';[4] the erotic is 'a relationship with alterity, with mystery—that is to say, with the future, with what, in a world where there is everything, is never there . . . with the very dimension of alterity'.[5] Relation with an other—communication—is relation with the future, an opening up to the possibility of the future as uncontrolled, uncontrollable, indefinable, unpossessable, the negation of what is. The efforts of administrative reason that are bent towards prediction, simulation, foreclosure, ownership of the future—the five-year plan, Rand Corporation and Trilateral Commission modelling, the strategic planning of arms and industry—are based on the erasure of mystery, the annihilation of the other, the extension of the same.

From this, I think it is clear why there is an ethical imperative towards communication, as well as the instinctive drive observable in our cities and our intercourse. This is ethics without morality: no knee bent to tradition or authority, but recognition that the very grounds of communication, the hope that it may produce something new, are today in play in the political economy of globalisation. To hold the ecological in mind: this openness is the grounds of evolution. Our task is to prise open the closing gateway through which the Messiah, the New Humanity, the next species might arrive.

It remains to speak of the aesthetic.

Aesthetic Futures

Beauty is not a quality of objects. The old saying is true, in this respect: beauty is in the eye of the beholder. We would have to say, in the body of the beholder, and to turn the language further to our purposes, to rephrase this as: beauty is the bodily experience of the other. There is something inherent in this otherness: an element of surprise, perhaps, but much more so a sense of opening oneself up to the experience, or sensing the invasion of self by some other, be it thought or sense experience. Today beauty is more valuable than it has ever been, not just because of its rarity, but because the hyperindividualism of neo-liberal capital, the narcissistic culture of consumption, militates against openness, and can only explain it in narratives of self-expression, confession and the healing of the self. Such localisation and consumption of the aesthetic misses entirely, and necessarily, the point: that to experience beauty is to experience the permeability of the skin, to feel viscerally the pull of other tides than those associated with the will.

Sometimes it is hard to distinguish the aesthetic from disgust: both are extreme states of the body, confronted with its unbounded connectivity with the whole world, synthetic and organic. The experience has nothing to do with taste, not even with the socialisation of taste, where groups communicate their solidarity with one another (and often enough their rejection of others) through demonstrations of shared cultural attitudes and practices. It has everything to do with the ethical imperative of communication: to enhance the communicative, so that every cell is engaged in communication. In this sense, the aesthetic is the highest expression of the ethical, impelling us towards, more than the recognition, the experience of interconnectedness. As such, it demands that we take responsibility for our experience, as it reveals the complete dependency of experience on the world. It occupies, for the materialist, the position that spirituality has for the religious, the place of hope.

The loss of subject in the aesthetic experience, or rather the irrelevance of subjectivity in the experience of connection, implies the irrelevance of objects, and more specifically of the subject–object relation. In turn, this implies the irrelevance of objects. Anything can be the matter of such experience: Proust's memory of the madeleine, Kristeva's horror at the skin on hot milk, the Shoemaker–Levy comet in the summer sky, my puppy leaping for a ball on the beach. But because we live in a period in which every effort of contemporary discourse is bent towards subject–object relations, there has to be an equal, indeed a greater effort towards making the aesthetic available. This effort, it seems to me, is the same as the effort to pry open the gates of the future. It is about the experience of alterity, of the availability, if not now then soon, of an utterly other mode of existence beyond the cultural imperative of selfish individualism.

In this sense, the ethical aesthetic presumes to question the value of art, even its very concept. This is, of course, a central function of art itself, probably since Duchamp; certainly since Kossuth.[6] What Kossuth retains, however, is the art-function, even as the art-object fades. In a sense, what there is of the aesthetic in art-objects is what is left when one has removed from Wagner his totalitarian pomp, or from Poussin his Jansenism. It is, perhaps, what one gains from the loss of familiarity: the shock of alterity that sometimes disengages itself from a Vermeer or a van der Weyden as the constitutive context dissolves, and the thing itself imposes itself as raw percept, as material. The absent artist has become future in their lack, and so achieved that utter otherness that makes them valuable in the current struggle. There are some works today that seem to operate in just this way: Rachel Whiteread's 'House' springs to mind.

Most art, however, is not aesthetic in this sense. It is, by and large, research into survival: how are we to retain some humanity in the face of what confronts us? Art, the mode of artisan production, the personal relations that the art world fosters, the scenting of coming storms, of the world or of the heart, acts without fundamental or radical challenge, simply because it accepts its status as art. As work, art offers one leading contribution: the example of creative labour. But in terms of its productions, it can only operate within the terms of art itself, a discourse in which the resilience of the dominant, its ability to contain and assume the resistant, has been apparent now for over a hundred years.

The challenge is to produce an art without artists, just as it is to produce media without professionals, and a polity without politicians—a public art, in which the private is dissolved, as it is in any case in the surveillance society, where the intimate and unconscious are leaking into the light; a public art that is more than the artist making their work public, or building works in public spaces, or directing the work of members of the public, but rather an art by, for, in the public sphere, art by virtue not of the success with which it moves expression from sender to receiver, but by virtue of how many people it inspires to communicate, and with what breadth and depth.

The new media may ossify in the forms that are standardised by cartel agreements in the interests of corporations now so huge and so removed from daily life that they should be considered as the first cyborgs, the first true hybrids of humans with information technologies. Alternatively, there is the possibility of creating networks of cultural translation, of mutual misunderstanding and confusion. The new-media arts, as is the job of art, are busy rejecting the Great White Future that has been offered to them, and have devoted themselves to producing a huge variety of alternative models, networks, software, concepts and modes of interaction that can have little or no use in administrative culture. Moreover, precisely because of their exclusion from the academies, they do so in the spirit of true amateurism: the act of making for the love of it—a love that itself is a relation with the other and the future. In doing so, they are rejecting the sense-making that is so characteristic of the managed present. In its place, effectively, they are making a kind of networked primal soup, that seething of molecules from which secular science suggests that life first evolved. It is in this sense too that we must be prepared to abandon the concept of art in favour of the concept of culture.

From such a mess of conflicting, contradictory, messy mediations, there is the possibility of new growth, unlike any we have seen or conceived of: precisely of that open future that Sardar demands of a refreshed futures studies. The work of media cultures is itself a form of advanced theoretical enquiry into the form and purposes of existing media, and the radical alternatives to them, and to art, even to theory and philosophy. They provide us with an alternative model of futures studies: a practical, hands-on materialism in the age of the simulacrum.

But an ethical aesthetics must confront the material realities of the world on which its work depends, and so give primacy to survival. Diversity is the principle of ecologies: diversity is the source of new evolutions. Survival is the first duty of the elements of an ecology, not only for their own sake, but in order to keep the broadest possible range of evolutionary pathways open. Standardisation and conglomeration are its enemies, and in Africa especially responsible for the death and exclusion of a huge sector of the world's population. Despite (or perhaps in the long run because of) their phenomenal current growth, the infotainment industries will waste away from lack of diversity, and will try again to ingratiate themselves with local cultures worldwide, in an attempt to turn them too into the familiar commodity form of capitalism. What I hope for from the new-media cultures is an alternative mode of access to the human community as the totality of relations between people, in which the machines and objects that today stand between us can become instruments of mediation, engines of possibility for an open and accessible future.

Notes

1. Manuel Castells, *The Information Age: Economy, Society and Culture*, vol. 1: *The Rise of the Network Society*; vol. 2: *The Power of Identity*; vol. 3: *End of Millennium*, Oxford: Blackwell, 1996–7, p. 98.
2. Webstalker <http://www.backspace.org/iod>
3. Walter Benjamin, 'The Author as Producer', in *Understanding Brecht*, translated by Anna Bostock, London: New Left Books, 1973, pp. 85–103 (p. 98).
4. Emmanuel Levinas, *Time and the Other*, translated by Richard A Cohen, in Sean Hand (ed.), *The Levinas Reader*, Oxford: Blackwell, 1989, pp. 37–58 (p. 44).
5. Ibid. 50.
6. Joseph Kossuth, 'Art After Philosophy', *Studio International*, 915–17: 134–7, 160–1, 212–13, October/November/December 1969.

15
Futures and Knowledge

VINAY LAL

Some ten years ago, when I was still a graduate student at the University of Chicago, I found to my considerable surprise that the influential journal *Alternatives*, which is jointly produced by the World Order Models Project in New York and the Centre for the Study of Developing Societies in New Delhi, had been shelved in the section reserved for journals in economics and operations research in the library of the university's business school. I could not have expected that *Alternatives*, which has featured much of the best work today on alternative conceptions of human needs, visions of human society, and modes of distributive justice, besides offering trenchant critiques of dominant models of war, violence, and political and economic repression, would be placed alongside journals in economics. Not everyone would necessarily agree that economics remains, singularly, the 'dismal science'; none the less, it is indisputably one of the most retrograde and parochial disciplines anywhere in the world, a discipline that, dare one say, will perforce have to be bludgeoned into humility before its perfidious pretensions to offering us reliable and worthy knowledge can be put to rest. In near proximity to *Alternatives*, on the other side of the shelf, lay various journals on forecasting, technological planning, and strategic management, as though the only conception of the future that we are permitted to have is one that the cormorant crew of economists, management specialists and technocrats, who have hitherto failed miserably in giving us a more desirable society, have ordained as worthy of the attention of humankind. Economists have flourished, just as the societies they have been called to manage have decayed; but unlike the tribes to which they are likened, economists have scarcely had the decency to live in self-sufficiency.

It is quite possible, of course, to describe the peculiar place that *Alternatives* occupies in the shelves of more than one university library as a technical problem in classification, or to attribute the error to the quirkiness of a few, perhaps ill-informed, librarians. All knowledge systems have relied on forms of classification, and the disposition towards one form of classification rather than another might be no less than the difference between competing visions of culture and society. Problems of classification and categorisation, as is quite transparent, are almost never mere trivialities: thus the history of Sikhs in post-independent India might have been quite different had not the British classified them as a 'martial race', nor would the Gurkhas have continued to do the dirty work for British imperialists had they not fallen under the same rubric; similarly, if the deaths on account of 'development' could be counted alongside the countless victims of numerous genocides, the 20th century might appear still more barbaric than it does even to the mere observer of history. It is through classification that 'Otherness' is marked, boundaries are demarcated, and entire cultures are construed as being 'civilised' or rendered as outside the pale of civilisation. The fate of those countries that the United States chooses to describe as 'outlaw' or 'rogue' states provides one of many instances of the power of naming and classification.

There is in the tale of the not-mislaid journal, then, a rather more ominous warning, about both the oppressiveness of modern knowledge systems and the manner in which, as Ziauddin Sardar has described it, 'the future has been colonised'. At a considerable distance in the past, the future was the provenance of astrologers, soothsayers, palmists and various other traditional specialists in magic, fortune-telling and curses. In a manner of speaking, however, every story-teller was a futurist, since stories (though they are located in the past) are invariably interventions in the future. Story-telling is especially associated with children merely because, as is commonly thought, children do not have a developed capacity for understanding abstractions, and consequently it is the concrete detail in which they are immersed. Everyone recognises that stories are vehicles for the transmission of moral lessons, but it is pre-eminently through stories that we convey to children our ideas of, and hopes for, the future: the future has no meaning without children, and it is in them that we invest our futures.

The ancient Greeks certainly recognised that no matter where one went, one was bound to encounter a story. It is no accident that the greatest of the Greek writers in the post-Homeric period was the boisterous and mythomaniac story-teller Herodotus; indeed, to take a heretical view, the demise of Greek civilisation can be marked by the advent of the historian Thucydides, who set out to correct the record, tame Herodotus's flights of imagination and present a more realistic account of Greek society. While the most outlandish and egregious of Herodotus's representations of the Other were absorbed over time into the West's huge corpus of ideas about purportedly barbaric, primitive or otherwise inferior and exotic civilisations, in every other respect Herodotus was sought to be disciplined. Herodotus's depictions of northern Africa as inhabited by headless or dog-headed people, with eyes in their breasts, or of black men as producers of black sperm, were considered as quite authentic representations of the Other,[1] but in other respects Thucydides' ruthless devotion to realpolitik and his rejection of sentimentality were seen as more reliable signposts to the future and the exercise of power. Western civilisation has ever since gradually been losing its capacity for story-telling, possessed as it is by the desire to scientise its narratives.

If at the lower end the astrologers and palmists reigned supreme, at the higher end the lot of thinking about the future fell to the utopian visionary and the prophet. Many utopian thinkers were, however, inclined to locate their utopia not in the future but in the past, in some imagined 'Golden Age' when law and order prevailed, and when justice was not so easily mocked. Though the tradition of utopian thinking survives in the 20th century, judging from the works of H. G. Wells, Eugene Zamiatin, Aldous Huxley, George Orwell and many other, lesser writers, it has been showing a precipitous decline for some time, and has now largely been relegated to the ranks of science fiction writers and their admirers, who are determined to establish that the American government has been conspiring to keep knowledge of Martians and other extra-terrestrials a secret.

In the United States, a country uniquely built on the promise of the future, nothing is as prized as the past, and most achievements are at once earmarked as 'historic': no utopian thinking can truly be contemplated among a people whose millenarianism is only an impoverished fundamentalism. Prophecy has been even more effectively pushed into complete extinction, asphyxiated on the one hand by the increasing dominance of the historical mode, and condemned on the other hand as a regrettable residue of medieval superstition, the remaining sibling of alchemy and black magic. In the English-speaking part at least of the Western world, Blake appears to be the last in the line of the prophets, but the entire West remains alienated from the prophetic mode, and not only because of the loss of orality, the transformation of the countryside, the overwhelming ascendancy of the print (and now visual) media, the declining emphasis on memory, the submission of civilisational entities to the nation-state and the disappearance of the classic itinerant. Though Marx and Freud might well be hailed as prophets in their own right, they remain resolutely the creatures of knowledge formations that envision no possibility for dissent other than in the language of those formations themselves. Thus Marxism can allow for no critique that is not historical, and indeed to be non-historical, or even a-historical, is to open oneself to the charge of belonging with the primitive, with those hordes still vegetating in the frozen vestibule of time.

Futures Studies?

It must come as a surprise to many, then, to find that the future is striking again. In the public domain, most particularly in the United States, the future is most often recalled in the conventional pieties of politicians' pronouncements, in their exhortation to us to remember what is good for 'our children's children'. It is to assure 'our children', and in turn 'their children', a bright future, free of biological weapons and poisonous gases, that President Clinton proposed to initiate the next round of the carpet-bombing of Iraq: if the preservation of the honour of English women was seen as conferring upon their menfolk the right to commit mayhem around the world, the invocation of 'our children's future' now similarly provides a sanctimonious license to discipline a recalcitrant world. More generally, however, the American tendency is to turn the future over to policy planners, management specialists, technocrats, and—most of all—computer whiz kids. In this vision, if so lofty a word may be used to describe the pedestrian rumblings of glorified plumbers, the World Wide Web and the Internet will keep us all connected, and 'chat rooms' and 'cyber cafés', we are induced to believe, will suitably substitute for table-talk and what once everyone understood as conversation. We may all be connected, but apropos Thoreau's comment upon learning of the invention of the telegraph, do we all have something to say to each other? While diagnosing the failures of the Englishman in the colonies, E. M. Forster—who divined as well the madness that men are capable of to preserve the sanctity of 'women and children' —came upon the sacred mantra to bring together the East and the West: 'Only connect.' But our modern form of connectedness is only a travesty of the feeling of community that now seems irretrievably lost, and a lesser degree of connectedness would do a great deal to render the world more pluralistic, more impermeable to the dominant categories of knowledge and homogenising contours of culture. Had we been more attentive to the political economy of hybridity, we might sooner have come to the recognition that multiculturalism has flowered in the most insipid, not to mention insidious, ways—in the elimination of multiplicity, and in the promotion of monoculturalism. We are connected, most surely, but only by the barest threads and lifelines that the dominant culture of the West deigns to place in other palms. The imperialists of a previous generation are the multiculturalists of our times.

No doubt, if the future was left to technoplanners and computer experts, we would achieve the same results envisioned by the creators of the neutron bomb, which while destroying all signs of life leaves buildings intact. Since the human being is the one unpredictable animal, many planners for the future find *Homo sapiens* to be a rather unpleasant reminder of the impossibility of a perfect blueprint. Yet, since the essence of the Western ethos is to strive (and not always, or even seldom, for the good), and merely to strive, mastery over the future—following the now-contested mastery over nature, women and children—is deemed an imperative. Consequently, like everything else, the future too has become a subject of study. Though the study of the future has only a fraction of the trappings associated with the traditional disciplines, future trends are there to be seen. The futurists have their own associations and organisations, their annual conventions, and their own organs of research and communication. Across the globe, especially in the 'advanced' Western nations, where the future will increasingly begin to look like their past, even a few university departments of future studies have cropped up. Though in some sectors the futurist has yet to gain academic respectability, since his or her calling is still associated with astrology, numerology, palmistry and other supposed superstitions, the study of the future is none the less now poised to become a big business. As the rest of the world embraces the market morality, and countries consent to the structural readjustment mandates of the IMF, the same breed that rules the roost in Western countries is beginning to flex its muscles in developing countries. In a country such as India, where business schools were something of an anomaly less than two decades ago, and where business had less than the tinge of respectability, no degree is now more coveted than the MBA. Suited and booted financial planners, consultants, management experts and computer specialists, who have learned the tools of their trade from the West (thereby enacting the modern form of Vishwakarma puja, a Hindu religious rite in which the worship of tools, which signify the Lord's creation, is undertaken), partake of the 'hotel lobby culture' described by Cornel West, and it is at their business seminars and lunches that they hatch those schemes designed to render the future of India (and much of the Third World) like the present of the West. Truly, the future of the greater part of the non-West, if the forecasters, planners and technoexperts are to be believed, is to be without any intimations of the future: it is to live someone else's life, to dream someone else's dream, to inhabit someone else's skin and to become someone else's merchandise.

In the matter of the future, one might then reasonably infer, it will be business as usual. For a very brief moment, it appeared as though this business would be fatally interrupted by the demise of communism in Eastern Europe and the Soviet Union. Some disciplines, such as anthropology, Oriental studies and historical studies, which served as the handmaiden of the West in its efforts to colonise the world, had as it seemed long ago outlived their usefulness; other disciplines, such as political science, which had flourished during the Cold War and had, in a substantial number of its practitioners, no other reason for being than to be the foot-soldiers of America's consumerist ambitions, political self-aggrandisement and ideological war on communism, should have faced extinction. But no species has ever willed its own destruction, however close human beings have been to eliminating themselves (and other species, who are not granted the dignity of survival in themselves, but only for humans); moreover, Western man knows little else if he does not know the art of re-tooling. All the questionable disciplines adroitly reinvented themselves and even became indispensable. It had been conventional to believe that only the West had history, but after the atrocities of the two world wars, and more particularly the genocidal impulses of the Germans, not to mention the American resort to nuclear terrorism in Hiroshima and Nagasaki and the Allied blitzkrieg in Dresden and elsewhere in Germany, much of this history seemed a bit unsavoury. If, after Hitler and Mussolini, the 'great man of history' theory no longer appeared so compelling and unambivalent, innumerable heroes could be recovered from the battlefields. Historical studies did not have to seek far and wide to find other hospitable homes: history was to become responsive by becoming the common man's chosen field, Everyman's hobby. Anthropology was supposedly to develop a critical bent, and become sensitive, self-reflexive and more pluralistic; this 'internal critique' continues, and has helped to shape, though not decisively yet, the future of anthropology departments, where the physical anthropologists and old cultural anthropologists have not quite exhausted their depredations. Oriental studies, meanwhile, had become reconceptualised as area studies, a matter even of vital national security, and to this have been added various forms of ethnic and minority studies. This form of self-fashioning, of acquiring an acquaintance with the Other, sampling various cuisines, and acquiring a library of 'world music', is now even championed as multiculturalism, as an instance of the West's unique thirst for knowledge and capacity for curiosity. It has almost nothing to do with the opening or closing of the so-called American mind.

What, then, of the political scientists, those members of the American academy who so shamelessly thrived on the foreign policy and defence establishments, or of the economists, who were so warmly embraced by the authoritarian regimes of Latin America or Asia? The latter, who have more lives than cats, have found their calling once again, as the former Soviet bloc opens itself up to the rapacious drives of the West, and the booming 'tiger' economies of Asia show unmistakable and alarming signs of weariness, poor management and what Western economists describe as an inadequate comprehension of the working of the 'invisible hand'. Nickel-and-dime capitalists once again can provide hoary testimony to the endurance of the American dream: the rags-to-riches narrative survives, though perhaps it is finding a more hospitable reception elsewhere: witness the ascendancy to the Presidency of India of a man from the class of dalits, once commonly described as 'untouchables', or the recent Prime Ministership of a man from a peasant community. Indeed, never has the economist had such a large playing field as he does now, when the entire developing world seems on the brink of 'liberalisation' and 'privatisation': the academic economist is easily transformed into a corporate economist. This apparent shrinking of the world, which we are told will make of the world a 'global village', is music to the ears of the capitalist and the economist alike. 'Globalisation' in effect means that the colonisation of the developing world, which in time was not even financially lucrative for European powers in some cases, can now be rendered complete. Moreover, whereas the exercise of power was once naked, and the victories of the battlefield were won by the ruthless bombing of villages, ironically termed 'pacification', and the Maxim gun, now domination will take place under the sweeter and gently killing dispensation of McDonalds and Coca-Cola. Cows must be fattened before they can slaughtered.

Clash of Civilisations!

But the story of globalisation scarcely ends here. Though the shape of the future under globalisation suggests unequivocally the narrowing of cultural options, the reduction of democracy to largely meaningless gestures at the electoral booth, the beggaring of the Third World and the instillation of the warped mentality of the West into people not so utterly incapable of dealing with the Other except by habitual recourse to various forms of total violence, what is most at stake is the future of knowledge itself, though the debate is most often cast in terms of 'culture'. The 'battle lines of the future', the political scientist Samuel Huntington argued a few years ago, will be the 'fault lines between civilization', and he sees the future as one in which the centre-stage will be occupied by a conflict between 'Western civilization' on the one hand and Chinese nationalists and 'Muslim fanatics', acting singly or in concert, on the other hand.[2] (It is axiomatic that 'Western' and 'civilisation' are supposed to be in natural apposition to each other, just as 'Muslim' and 'fanatics' are regarded as making happy bed-fellows. Perhaps we ought to place 'Western' and 'fanatics' in apposition to each other, and the two together in opposition to 'Muslim civilisation'.) That this thesis should have been advanced by 'mad-dog Huntington', as he was known to radicals in the 1960s, is scarcely surprising. First an avid proponent of America's involvement in the Vietnam war, and then an advocate of the nuking of Vietnam, Huntington became one of the foremost ideologues of the Cold War and was subsequently to find himself much in demand as a mercenary, offering such advice to despotic regimes as would enable them to be properly authoritarian, all to equip themselves for a future of democracy. Since the end of the Cold War threatened to run him out of business, leaving him with a mere Harvard sinecure, Huntington had to reinvent himself, and rather adeptly—as the millennium draws to a close—he chose a Spenglerian worldview to paint the picture of the future.

Huntington's thesis is all too simple: where previously world conflicts were largely political or economic, the new conflicts will be largely cultural. Having identified seven or eight civilisations—'Western, Confucian, Japanese, Islamic, Hindu, Slavic-Orthodox, Latin American and "possibly" African', one suspects in approximate descending order of importance and worthiness—Huntington advances the view that these civilisations are bound to be in conflict with each other. 'Western ideas of individualism, liberalism, constitutionalism, human rights, equality, liberty, the rule of law, democracy, free markets, [and] the separation of church and state, often have little resonance' in other cultures. These differences, 'the product of centuries', will endure; they are more lasting than 'differences among political ideologies and political regimes'. Many developed countries modernised, introducing such technology and administrative efficiency as would enable them to raise the standards of living for their people and compete in the world market; but these countries did not Westernise, as it was scarcely to be expected that they would compromise, for example, their collectivist spirit (as in the case of China and Japan) for the individualism that is ingrained in the Western psyche. Huntington takes it as axiomatic that these alleged differences will lead to conflict, which he sees as exacerbated by the increasing tendency towards 'economic regionalism'. Many of these civilisations, in other words, are developing, or have developed, their own trading blocs; and the increasing worldwide trade is not so much between different trading blocs, as it is within these blocs. Thus, howsoever inadvertently, Huntington returns to the thesis that world conflict will be bound up with economic competition, as though numerous forecasters have not been labouring to make this transparent for a very long time. (One discerns even, in some of these prognoses, the desire that trade wars should lead to something more manly and rougher. War has been good to the West: if Germany was the classic case of a nation built on a war machine in the first half of this century, it is the armaments industry and the 'military–industrial complex' that powered the United States to the status of the world's pre-eminent power in our times.) The attempts of the West to impose its values upon other cultures are countered by other civilisations, Huntington argues, and as non-Western civilisations are no longer mere objects but 'movers and shapers of history', violent resistance is to be expected. The clash of civilisations 'will dominate global politics' and determine the shape of the future.

Though Huntington already finds such clashes between civilisations taking place around the world, the brunt of his thesis is that the West should expect conflict between itself and the two entities, China and the Muslim world, that most have the power to resist the West's continuing influence. Islam is most hostile to everything from which the West draws its sustenance, and drawing upon an earlier and lesser-known piece by his fellow Ivy-League cohort, the Orientalist Bernard Lewis, Huntington is inclined to believe that Muslims can tolerate neither the separation of Church and state, nor submission to infidels. It is not that imperialism and domination are unacceptable in themselves to Muslims: rather, as Lewis put it, 'What is truly evil and unacceptable [to Muslims] is the domination of infidels over true believers.'[3] In Huntington's cryptic formulation, 'Islam has bloody borders'. While the 'roots of Muslim rage' are an ambition that is thwarted and the paramountcy of Christianity over Islam, it is, on the other hand, the insularity, arrogance and self-centredness of China, which is now poised to rule over the Pacific, and where since time immemorial totalitarian regimes have crushed the rights of the people, that make it an implacable foe of Western democracies.

While Huntington and others of his ilk have never been known for the rigour of their thought, and it would be fanciful to expect them to be conversant with the theoretical movements that have over the last two decades greatly aided in analysing political and cultural discourse, it is evident that Huntington is unable to make the most elementary distinctions between the nation-state and civilisation, or understand the consequences of injecting essences into politics. So the dispute over Kashmir becomes part of the fabric of 'the historic clash between Muslim and Hindu in the subcontinent', and since 'Hindu' and 'Islam' point to two kinds of civilisations, a conflict between nations-state, themselves the product of forms of reification of identity under the oppressive political paramountcy of a colonial power, becomes a conflict between civilisations. Just how 'historic' is the 'historic clash' between Islam and Hinduism on the Indian sub-continent, and does this history take us back to 1947, to the early part of the 19th century and the inauguration of what the historian Gyan Pandey has described as the 'communal riot narrative', or to the beginnings of Islam (as Huntington would no doubt like to believe) in India in the 8th century? Huntington recognises no 'Indian' civilisation, which has been infinitely more pluralistic than anything European Christendom has ever known, just as he transposes the experience of Europe's bloody religious past on to every other place, assuming of course that there could have been no superior form of social organisation, and more elastic conception of self, outside

Europe. Huntington's view here is the primitive one that religion must remain the fault line between civilisations, because howsoever insistent the West may be on retaining the separation between Church and state, other civilisations have their essential and inescapable grounding in religion. Though the West orients itself spatially, the Orient renounces spatiality: thus we have Confucian but not Chinese civilisation, Hindu but not Indian civilisation, and Slavic-Orthodox but not Eurasian civilisation. Huntington's essences are the stuff from which the proponents of the national-character industry made their living and killing some decades ago.

If Huntington fatally substitutes 'Hindu' for 'Indian', no less egregious is it to suppose that the West is the repository of such ideas as 'human rights' and 'the rule of law'. This leads us, of course, to such nauseating spectacles as the report of Madeleine Albright, the US Secretary of State, lecturing the Vietnamese about their inadequate respect for human rights, which the United States did everything to decimate in Vietnam and has done little to respect around the rest of the world. After having engineered the first genocides of the modern period, and being responsible for the most gruesome ones of our own bloody century, the West can applaud itself for having successfully induced some other peoples to act in its mirror image. Since the hypocrisy of the West in these matters is now too well known and documented to require any further comment, a more disturbing and less-noticed strategy to colonise the future of much of the world deserves scrutiny. As the record of the West—after the overwhelming evidence proffered by the genocidal annihilation of native populations in the Americas and Australia, the history of colonialism around the world, and the Holocaust within Europe itself—in perpetuating unspeakable atrocities can no longer be denied or camouflaged, a species of American or Western 'exceptionalism' is introduced to differentiate the actions of the West from those of the rest of the world. 'The accusations' against the West 'are familiar', concedes Bernard Lewis, and to them, he continues, 'we have no option but to plead guilty—not as Americans, nor yet as Westerners, but simply as human beings, as members of the human race.' Since this is too prosaic for words, Lewis moves to a higher level of comparison: so the treatment of women, deplorable as it has been in the West, 'unequal and often oppressive', is none the less said 'at its worst' to be 'rather better than the rule of polygamy and concubinage that has otherwise been the almost universal lot of womankind on this planet'. But as even this is not wholly convincing, and certainly subject to dispute, Lewis moves on to what he imagines is the loftier form of the argument. What is 'peculiar' about the 'peculiar institution' of slavery is that in the United States it was eventually abolished—and so from here impeccably to the *non sequitur* that Westerners 'were the first to break the consensus of acceptance and to outlaw slavery, first at home, then in the other territories they controlled, and finally wherever in the world they were able to exercise power or influence—in a word, by means of imperialism'. Having first introduced slavery to many parts of the world where it was never practised, Western powers conspired, against their own self-interest, to outlaw it: Lewis finds that a charming thought. So the West is 'distinct from all other civilizations in having recognized, named, and tried, not entirely without success, to remedy those historic diseases' such as racism, sexism and slavery: herein is the exceptionalism of the West.

While the general tenor of American or Western exceptionalism similarly defines Huntington's enterprise, he follows the complementary strategy of demonstrating the ills that must result when the rest of the world emulates the West. It is not that the West must not, in principle, be emulated: we should all aspire to the greater good. Moreover, since the natives thrive on mimicry, a good course must be left for the non-West to follow. If there is to be a universalism, it can only be universalism on the Western model; but what sort of universalism is it that makes the world capable of equating America not with democracy, the Bill of Rights, the spirit of freedom and the inalienable right to the pursuits of happiness and liberty, but with Pepsi, Madonna and McDonalds, in short with 'US pop culture and consumer goods'? 'The essence of Western culture', Huntington is at pains to persuade, 'is the Magna Carta, not the Magna Mac', the English of Shakespeare and not of pure instrumentality, and the conflation between the two cheapens the inestimable achievements of the West.[4] Nor is this all: if the West can even turn its evils into good, as Bernard Lewis pleads apropos imperialism, Huntington inclines to the formulation that the rest of the world habitually renders all that is good into evil. Most notably, Huntington sees this in the trend to accept modernisation but not Westernisation; and if this appears to be merely innocuous, or at most mildly self-serving, consider that 'when non-Western societies adopt Western-style elections, democracy often brings to power anti-Western political movements'. The warped minds of tin-pot Asian and African despots render democracy instrumental to totalitarian impulses and designs: 'Democracy tends to make a society more parochial, not more cosmopolitan.' It must perforce, on Huntington's view, be Western exceptionalism, rather than Western universalism, that will mark the future.

It is an extraordinarily telling comment on the state of knowledge and the academy that so poorly conceived, not to mention absurdly reductionist, arguments should have received the accolades that Huntington's papers have garnered. If the future is to be defined between the polarities of Western exceptionalism and Western universalism, then the only future that remains is to opt out of this debilitating and diseased view of the future. This is less than easily attained, since the West has even attempted to foreclose all dissenting futures. No dissent that does not take place in a form understandable to the West, or according to its canons of civility, is constituted as dissent; and every act of dissent that calls into question the purportedly dissenting frameworks of knowledge thrown up by the West in recent years, whether encapsulated under the term 'post-colonial' or 'postmodern', is construed as a retreat into romanticism, indigenism, nativism or tribalism. As our very idea of the future has been held subject to the dominant ideas purveyed by the experts, it becomes inescapably clear that no better future can be expected unless we can decolonise the dominant framework of knowledge.[5] The 20th-century university, with its disciplinary formations and its well-intentioned multicultural brigades, will have to number among the first of the victims. The earliest measure of our new-found wisdom and knowledge might well be to recognise that when the experts are removed from their area of expertise, the future will begin to manifest itself as more ecological, multifarious and just.

Notes

1. V. Y. Mudimbe, *The Idea of Africa*, Bloomington and London: Indiana University Press and James Currey, 1994.
2. Samuel Huntington, 'The Clash of Civilizations?', *Foreign Affairs*, 72(3): 22–49, summer 1993.
3. Bernard Lewis, 'The Roots of Muslim Rage', *Atlantic Monthly*, September 1990: 47–54.
4. Samuel Huntington, 'The West and the Rest', *Prospect*, February 1997: 34–9.
5. Vinay Lal, 'Discipline and Authority: Some Notes on Future Histories and Epistemologies of India', *Futures*, 29(10): 985–1000, December 1997.

16
Futures and Prophecies

JEROME RAVETZ

Futures studies has become an instrument for the colonisation of the future. As a field, it has become obsessed with preserving the 'power and territory' of the Western civilisation and the white, Anglo, middle-class professional futurists. So we read in the problem statement that opens this volume. The issues of power and territory are an exemplification of the social relations of knowledge under the conditions of its commodification. That is, unlike in the old days when knowledge was generally cultivated on behalf of an elite that commanded power and high culture together, now knowledge must earn its living on the market-place. For this it requires legitimation in secular institutions (based on universities) and also patrons, who are now quite utilitarian in their outlook. For the former, the apparatus of learned journals is essential; for the latter, corporate clients must be given the promise of something that they believe they need. Otherwise the field will languish, become obscure and underfunded, and relegated to the outer margins of the world of the intellect, attracting only zealots and cranks to its pursuit.

Nowhere do these tendencies operate more strongly than in the USA; and given the size and strength of the US knowledge industry, they will inevitably operate worldwide. In both respects, they will enforce a narrowness and short-sightedness of the vision of the field. It may seem paradoxical or even contradictory that 'futures' should succumb to the same vices as, say, neo-classical economics or 'decision theory'; but the iron laws of knowledge economics permit no easy exceptions. For the standard machinery of production and validation of knowledge is essentially elitist and hegemonic—this has been well noted even in the areas of the natural sciences, where the materials are at least ostensibly free of cultural or political bias.[1] Also, the field's corporate clients, not reckoning 'futures' as 'culture', demand a product and activity that they can understand; and hence it will reflect the vulgarity that is dominant in that milieu.

Of course, it was not always that way. There was a time when futures may have seemed quite radical, when the concerns of the 1960s, with environmentalism and consciousness to the fore, temporarily displaced corporate theorising from the public centre of the field. But that uneasy mixture could not survive, except perhaps in the memories of particular practitioners. With the decline of the utopian thrust of the 1960s, futures had to return to its old client-base, or wither. Then it would necessarily follow fashion: globally aware in the 1970s, and less so in the

1980s. And all that time, it was maturing itself in the conventional academic way, so that there would, by the 1990s, be available a solid corpus of writings about writings and materials for teaching.

These are now offered to the wider world, in a fashion that the proponents of the colonising thesis find objectionable, since it is so hegemonic and destructive of non-Western endeavours in the same direction. But what else could be expected? What alternative coherent vision of the future is there, especially now that socialism is discredited, and the various renovated traditionalisms are still in an early stage of struggling for an identity? It could be that the various radical and futuristic Third World groups might do best simply to borrow the enemy's weapons and use them creatively, as Sardar has shown recently in his profound essay on the future influence of CD-ROM technology on Islam.[2]

I find a similar inevitability in the other object of Sardar's strictures, that is, the 'More Syndrome'. To be sure, the various appropriators of 'Eastern wisdom' may well have a scant competence in that area, and may purvey deeply misunderstood versions of its teaching—as Sardar demonstrates so powerfully in *Postmodernism and the Other*.[3] But when has it been different? A culture can appropriate only what it understands in terms of its own needs. Fritjof Capra may well have been wanting to integrate his own inner experience, explicable only in 'Eastern' terms, with the public experience of the West, so as to help the West. Translating it into four-dimensional physics is doubtless a betrayal; but otherwise he would have to follow Ram Dass and the other converted intellectuals into the cultural wilderness. And in the East itself, what became of the heretical Hindu teachings of The Enlightened One as they travelled north-eastwards? In Tibet they were assimilated to the local spirit worship; in China, they were absorbed into the veneration of the androgynous god/dess Gwan Yin; and finally in Japan, they were put to the service of a warrior caste.

It is always thus. Rather like Dr Johnson's woman preacher, the wonder is not that it is done badly (as the dog dancing on its hind legs) but rather that the journey to the East has happened at all. In cross-cultural terms, we may say that the terms of exchange between West and East, for so long dominated by the force of Western material technology, have suddenly experienced a partial reversal. The technology of consciousness, previously quite foreign to mainstream European culture (even to most of its mystical strands), suddenly became a mass commodity, in the cultural, material and even political spheres. Just five centuries lay between the onset of European expansion and the 'Great War' and now, nearly a century on, we find prospects for a future in which European/American hegemony becomes increasingly dubious.

What we have in Sardar's statement of the problem is a case of a field now arriving at the state of a matured, 'normal' science in Kuhn's terms; but one where, because of the largely external clientele, the paradigm must be carefully neutered so as not to cause offence. Really, it was nearly always thus; and the elaborate studies with Delphi techniques, from the earliest days, always demonstrated that the future would not be too uncomfortably different from the present. But such 'normality' has its own dangers for the field. At the dawn of the 21st century, the most prudent assumption seems to be that all bets are off. Will the former Soviet Union go the way of Yugoslavia? Will AIDS become a worldwide pandemic, the postmodern high-tech successor to the plague? Will the Pacific Rim countries recover from their economic setbacks and continue, through innovation and management, their reverse-colonisation of key nations in the West? Indeed, in the USA itself, to what extent will the white-male intellectual hegemony survive the upsurges of non-Western groups, with (for example) Hispanics grabbing political power, Asians occupying the elite education, and all sorts of people demanding cultural validation in the core curriculum?

Questions such as these are not intended to be definitive. For me, they indicate how very different the near future may well be. A futures studies dominated by scholars trained and conditioned to concentrate on the continuity of a certain cultural and political hegemony might well find itself a quaint survival when the future arrives. The most prudent assumption seems to be that all bets are off. Just before I wrote these words there came the Official Denial that the asteroid of 2028 will collide with the Earth. What better recipe for total unknowability?

Indeed, as we watch the present move into the future by slides and bumps, what is impressive is how little we know, even of tomorrow. There may have been some smart money that sized up most of the South East Asia economies as kleptocracies heading for the crunch rather than as rapidly developing economies 'with an improving human rights record'. But it wasn't much in evidence, either to the public or even to the big boys.

Somewhat naively I (along with others) might ask, how can we talk sensibly or usefully about conditions in the medium or distant future, when we don't even know what's going to hit us next week? Although it might run counter to the professional in futures studies, we might think of the endeavour in terms of a play on the meaning of 'prophecy'. Only now is this interpreted simply as dealing with the future. Originally, to 'prophesy' meant to speak as inspired; the utterances, usually as warnings, would necessarily be largely in the future and conditional tenses.

In those terms, 'futures' would be analogous in its function to 'science fiction'. In that case, prophecy is framed (or thinly disguised) as literature. In this, its format is quasi-academic. The quality of any particular contribution would then be judged by its integrity, coherence and impact, rather than on the length of its list of references. Of course, prophecy has always been a risky game, both intellectually and personally; and this reflection of mine is by no means an exhortation for all readers to go out and make themselves noisily unpopular. Rather, it is a case of adopting a mode of discourse more appropriate to the function of warning and awakening.

For example, we might get out of the habit of treating possible futures as extrapolations of present trends and then being surprised when they are not. There have been many discontinuities in recent history, some of which were observed and analysed as they unfolded. For those, it is useful to think of what might (following

Mao) be called 'leading contradictions'. These are problems that the system cannot solve in its own terms, and that are sufficiently critical that if they are not 'resolved' somehow or other they could destroy the system. Thus, for example, the problem of 'distribution' was central to European industrial society; analysed by Marx as 'expropriation', it seemed for a while to threaten the foundations of the social order. But then it was resolved, partly by advances in technology but also partly by being exported to the poor countries and the natural environment. For the contradictions of state socialism, a system forged in the relatively primitive construction of industrial economies but incapable of guiding a matured consumer society, there was no resolution of the contradictions, nowhere to go, and so it stopped.

Any discussion of 'leading contradictions' is bound to be somewhat speculative, but it can focus our attention on the sorts of things that are likely to produce discontinuous change, especially the unwelcome sort. For example, the French depend for their energy partly on an ageing set of nuclear reactors of a single design, and partly on an ageing kleptocracy that becomes increasingly tyrannical. The State of Israel exists as the result of tragic exigencies that converted a messianic, otherworldly dream into the rationalisation of a warrior state; and where the self-appointed prophets of the dream have created a nightmare for conquered people and co-citizens alike. The United States of America depends on its founding ideals of freedom and equality, but now pays the price for the multiple crimes (genocide of natives, enslavement of Africans, rape of a continent) through which its prosperity and promise were achieved. No one who visits the vast areas of devastation and hopelessness of America's major cities can believe that this nation has solved its contradictions, or is immune to some future great catastrophe.

We might even define 'underdevelopment' as the state where the destructive contradictions are not being resolved, and are not even masked. This condition manifests as what we call 'corruption' and in tendencies to a Hobbesian 'state of Nature', involving the oppression of all who are vulnerable in any way by reason of ethnicity, gender, belief or powerlessness. It is of course possible that within such societies there will be sectors that 'grow' or 'develop' towards the ruling Occidental model. Indeed, the ruling liberal myth is that such is the natural path for all the less-favoured nations. To question or deny the force of that myth is to involve oneself in the contemplation of terrible, catastrophic tragedies; for that is the course on which so much of mankind is embarked. But to acquiesce in the myth, once one is aware of contradictions as the driving force of social change, is to accomplish a sort of lobotomy on oneself.

When we think in dialectical rather than linear terms, we are better equipped to consider the future in terms that open up basic questions about our existence. In Western thought these are described as 'eschatology and theodicy': why are we here, and if there is an answer in terms of a good purpose, why is it so regularly betrayed? The theological framework for such questions is no longer popular, and so they must be given a secular form. One such might be in terms of group identity. Looking around the world, we might ask, what group (national or ethnic) does *not* have a problem with its identity, does *not* feel that it is some sort of Other in relation to an Authentic group? Traditionally, it was the English, the Americans, and perhaps the French, who were in that unproblematic self-consciousness—perhaps because they were dominant powers. Otherwise, every group has struggled with existential angst of some sort or other, of lesser or greater severity and destructiveness.

Of course, those born to privilege in any society will tend to feel greater security, even if their larger identity is problematic. But even they pay the price for broader identity, in sharing the insecurities of their less fortunate brethren. Sometimes insecurity and arrogance seem to co-exist, but that might be a manifestation of the brutality whereby self-negation is projected outwards. And the condition of self-confidence is itself unstable; it was just a year after the Diamond Jubilee that the British bungled the Boer War, and the words of Kipling's prophetic 'Recessional' began to come true. Then came Flanders Fields and the squandering of the nation's manhood; and it was never the same again. For the Americans, the ending of spiritual virginity came in the duo-decade from the early 1960s to the mid-1970s, which began with one President assassinated, ended with another disgraced, and in between endured assassinations (Martin Luther King Jr. and Bobby Kennedy), bloody urban riots, the counterculture (psychedelic and political) and the trauma of fighting and losing the wrong war. Ever since then America has hovered between nostalgia and fear.

My point here is not to create a taxonomy of ethnic security, but to provide some perspective on insecurity. My point may not even be politically correct in effect if not in intent; it may be that all the insecure ones need to believe in the uniqueness of their suffering in order to give it meaning. Thus the Jews tend to see the Holocaust as theirs alone, being embarrassed to find comrades among Gays, Gypsies and even Slavs. And each victim-nation of past imperialism has a folk memory of it being an absolute evil, uprooting some previous happy state. There is then a danger that remembered pain will only feed self-pity, and will not show the way to compassion and to forgiveness—of oneself for victimhood as much as of the aggressor. It is then all too easy for yesteryear's victim to become today's oppressor, and to practice dehumanisation on those who are inconvenient to its present project. Any qualms are suppressed by the folk memories of the unique experience of suffering, which then justifies all present and future actions.

Is there a moral in my observation of this sad business? Perhaps it is that, as Gandhi said, 'non-violence' is the path requiring the greatest strength. It is also a source of growth, of healing, and of power. It is only 'truth and reconciliation' that gives South Africa some hope of survival; and similarly Northern Ireland. The alternative has now been defined quite starkly: the 'Kosovo-Jerusalem syndrome'—my ancient pain must now be alleviated by your future suffering.

The other theme I wish to develop deals with 'theodicy' and with the prophet's role. Looking at the world around us, we find evil popping up everywhere. Sometimes it is rampant, destroying lives and values on a massive scale, and leaving broken and tormented victims all over. And then a look at human history, at least from the rise of civilisation, shows how war and oppression have been the universal norm up until very recently indeed. Those who discover some particular evil for the first time cannot bear that pain, and so they dedicate themselves to stamping it out, while other, perhaps worse evils, flourish all around them. After some years or decades of action and reflection, some may come to see that simple exposures and denunciations, however necessary, are not sufficient. And worse, simplistic solutions, based on simplistic analyses of the causes of the evil, can produce worse evils than those they replace. It seems that uncontrolled power wrapped in moral rhetoric can be even worse than simple naked power. Hence again, the sort of dialectical thinking whose outcome is 'non-violence' provides (for me) the only hope that things can change for the better.

The history of the past few centuries has been a sort of experiment in the Enlightenment vision that material progress would provide a foundation for moral advance. In many ways it has done so, quite genuinely; but its old contradictions (e.g. imperialism) have never been fully resolved, and its new contradictions (degradation of the natural and spiritual environment) become ever more acute. Only with dialectical thinking, and compassion for those caught up in evil structures and ways, can we move on.

This conception of prophecy is very different from that ascribed to the greater prophets of the Hebrews. For them, it was denunciation all the way, with multiple dire warnings and lamentations. Only among the later, 'minor' prophets do we get the visions of each man sitting under his fig tree, and of the lion lying down with the lamb. We should not imagine that they got to this vision the easy way, by wanting everything to be nice for everybody. But they had seen the theocratic power-game played out for some centuries, and this was their response.

Such a vision is not at all easily translated into straightforward practice. Mass movements need simple, even simplistic slogans; and even Gandhi was followed by Nehru, just as St Francis was followed by Brother Elias. But we have gone through a century's worth of mass movements, promoting every kind of 'ism', and leaving us with a hideous mess. Perhaps those of us who can might give dialectics a chance. Following the thought of the Chinese sage, we might try to do less, and see whether that actually accomplishes more.

Notes

1. See Jerome R. Ravetz, *Scientific Knowledge and Its Social Problems*, Oxford: Oxford University Press, 1971; New Brunswick: Transaction Publishers, 1996; Jerome R. Ravetz, *The Merger of Knowledge with Power*, London: Mansell, 1990.
2. Ziauddin Sardar, 'Paper, Printing and Compact Disks: The Making and Unmaking of Islamic Culture', *Media, Culture and Society*, 15: 43–59, 1993.
3. Ziauddin Sardar, *Postmodernism and the Other: The New Imperialism of Western Culture*, London: Pluto, 1998.

17
Futures and Dissent

ASHIS NANDY

It is astonishing how some concepts, words or phrases acquire a life of their own in the public mind. After a while, they begin to look like axiomatic truths that the earlier generations did not know, because they were either ignorant or simply stupid. However, the life of a concept is brief in the social sciences and humanities; after a while, its clay feet begin to show and it begins to lose its glitter. Many begin to suspect that the axiomatic status of the concept is at best a minor irritation, at worst a sanction for new forms of violence or expropriation. Of course, the larger ideological frame—and, more than that, the emotional pattern—underpinning the concept does not die that easily. In fact, it might show more resilience than the concept itself. After a while, while social knowledge and intellectual sensitivities begin to move in one direction, the ideas and emotions that contextualise political choices, national policies and newspaper editorials move in another.

Thus, 19th-century social evolutionism, enthusiastically exported to the colonies by a galaxy of European social and political thinkers and gratefully received by their local disciples, precipitated a wide search for an open, uncritical Westernisation in much of the Southern world. Gradually, as the colonial political economy consolidated its hold in large parts of the world—initiating the first phase of the globalisation at a time when the term had not yet become a buzz word—many took it for granted that all non-Western societies would, in the long run, have to approximate the ruling societies of the West. It became a part of the structure of global commonsense—sustained by a new middle class trying to share and acquire the trappings of the new global culture of 'cosmopolitanism'—that ultimately every society would have to be 'self-critical' and re-examine what ought to be retained from its traditions and what ought to be jettisoned. This self-examination, of course, was to be conducted not from the point of view of traditions or those living with traditions or, for that matter, from a 'culture-free' or 'culture-fair' vantage ground (as psychologists define the theoretical posture), but from within a framework of absolutised universal values. It meant rating cultures like recalcitrant job applicants from a 'culture-free', 'rational', 'scientific', aggressively 'modern' baseline for their good and bad points. In practice, it also meant nothing less than waging an uncompromising battle against anything that went against the newly internalised, now-clichéd vision of a desirable society that emerged directly out of the European Enlightenment. Even many traditional scholars and theologians began to talk of re-tooling their own skills and conceptual frameworks to fit the emerging global market of ideas.

Gradually, in this century that kind of Westernisation fell into disrepute. I have lived through those transitional times and met in my childhood and youth hundreds of people who considered Anglicisation to be a dirty word and hundreds of others who felt proud when told that they were Anglicised. Actually, till the mid-1950s, very few people talked of Westernisation; 'Anglicisation' was the common expression, though some called it 'Europeanisation'. Over the decades, before my eyes, Anglicisation went out of fashion and people began to talk of Westernisation as if it was a new concept. Anglicisation was seen as colonial; Westernisation as something intrinsically good, something that involved the acquisition of cultural skills associated not merely with the European imperial powers but with Western societies in general.

But even Westernisation after a time lost its shine; it became all too obvious that the process presumed an inescapable hierarchy of cultures and nations. Specially, after the advent of the structural-functional schools in the social sciences, looking for general theories of social change, Westernisation sounded blatantly parochial or, as the then-popular expression went, ethnocentric. Following the aptly named Professor Parsons and his numerous apprentices speaking like noviciates in a new priestly order, we were taught the beauties of modernisation. It not merely sounded technical but also conveyed its linkage with the Enlightenment vision of the universality of social knowledge. Everyone suspected that between the old-fashioned concept of Anglicisation or Westernisation and the new idea of modernisation, there was no substantive difference, but few had the courage to say so in polite society.

The career of modernisation did not turn out to be a happy one either. The disgruntled would always be disgruntled, and many began to find faults with modernisation, too. Gradually, the critics of modernisation even began to outnumber the defenders of modernisation in many areas of knowledge. Even the emergence of postmodernism, which can be read as a dissenting child of modernity pretending to be an orphan, has made it unfashionable to talk kindly about modernity. Modernisation has been increasingly left to government bureaucracy, journalists, economic policy-makers, industrial tycoons, and the superannuated sections of the Left. Once again, the cutting edges of knowledge and intellectual sensitivity have become divergent from the world of policy and decision-making.

Within this intellectual climate has entered the new idea of globalisation, with its less-than-innocent presumption that the earlier systems were not global at all. As if, from Pax Britannica to Pax Americana, it had all been a movement from one kind of the local to another. As if colonialism was not a global phenomenon and did not endorse a particular kind of global political economy. As if the post-world-war era did not have a global cultural order, and only the two world wars were global.[1]

It is in this context that I shall summarise the thrust of the 'global consciousness' that is unfolding in front of our bleary eyes as we hand over intellectual responsibilities to a new generation of young South Asians.

The world 17th-century Europe created depended on three crucial innovations: (1) the modern nation-state system as it was 'formalised' by the Treaty of Westphalia in 1648; (2) the theory of progress, including the various editions of social evolutionism, as the principal way of conceptualising social time; and above all (3) the idea of modern scientific rationality as the ultimate organising principle and source of legitimacy in the modern society. All three were imposed through the new kind of colonialism that emerged in the 19th century, a colonialism that systematically laid down the ideological basis not so much of conformism as of dissent. Dissent now had to conform to the tenets of 'sanity', 'rationality' and 'progress'—all defined strictly according to the European Enlightenment. So much so that generations of Southern scholars would spend their lives discovering or researching the existence of the same elements in their own cultures (as historian and philosopher of science Deviprasad Chattopadhyay and economist–philosopher Amartya Sen have done, or are still doing).[2]

There were other innovations of the Enlightenment that were as important. Secularism and nationalism come to one's mind immediately. But the three listed above are the ones that have acquired axiomatic status over the last three hundred years. The others, such as secularism and nationalism, often derive their legitimacy from the three. Taken together, the troika can be seen as markers of a sanity that defines the limits not only of conformity but also of dissent. All dissent, to even qualify as dissent in the civilised world, has to meet the criteria set by the Enlightenment vision of a sane society and true knowledge, and, even more important, they have to use methods modelled on those pioneered by Baconian science.

Times change. It is becoming obvious to many at the end of the millennium that the three criteria that sought to define the limits of social criticism have themselves become major candidates for criticism. For these criteria in the meanwhile have become the legitimising principles for new kinds of violence, authoritarianism and exploitation. Human beings, one suspects, are psychologically prone to subvert all theories of liberation and convert them into new orthodoxies available for easy use by the privileged and the powerful. This has changed the context within which different schools of social criticism have operated during the last century. What seemed sane and realistic at one time increasingly look like instances of crackpot sanity and a highly romantic idea of realism having little to do with the world in which we live.

Thus, many in the Left, offering docile, in-house dissent from within modernity, have always focused on the faulty social relationships of modernity. The radical theories of secular salvation, too, have been eager to recover the pristine aesthetics of the core concepts of the Enlightenment, reportedly distorted temporarily by human irrationality, greed and cognitive failures, particularly in the Southern world. These internal criticisms have no awareness that large tracts of the non-Western world have begun to live in a world of transient theories of oppression and liberation. For it is now pretty obvious that human beings, given plenty of time and a long enough rope, can turn any theory of liberation into a new justification for violence and exploitation. This century has taught us that it is irrelevant whether or not the modern state has really been less expropriatory or violent than the *ancien régime*, whether or not Francis Bacon's objectifying, masculine science has really been an improvement on the vitalistic sciences that depended on what Sigmund Freud might have called projective identification with nature, whether or not the theories of progress—so lovingly developed by generations of thinkers for the benefit of the ill-fed, ill-clad and ill-educated in the South—have really ensured a more self-confident, optimistic world at a time when the older certitudes continue to dissolve.

The fact remains that each of the innovations introduced by the European Enlightenment as a principle of social organisation has become a new justification for violence and expropriation. This is so at a time when such violence and expropriation are now no longer that easy to justify in terms of principles once globally popular—religion, race, superstition or stereotype, loyalty to a local or national potentate, family honour, interests of a tribe or a clan, and so on.

The return to traditions in the Southern world in recent years, including what may look like a nostalgic invocation of the past, must be recognised as the marker of the beginning of an era when the Enlightenment vision, after more than two centuries of hegemony, is losing its sacred status. Though such returns also often produce psychopathological responses in the guise of revivalist or millennial movements, they must also be seen the same way as we have learned to see the Hellenic sciences at the beginning of the age of modern science. Everyone knows that there were other traditions of Hellenic sciences (the Pythagorian tradition for instance) that were ignored by the partisans of modern science, but one also knows that the return to Hellenic sciences by these partisans was not a return to true or historical Greece; it was an attempt to distance oneself from medieval science and deny any continuity between medieval and post-medieval systems of knowledge. Nor were such partisans of modern science terribly perturbed by the other kinds of persisting or recovered paganism that found expression in various forms of folk 'superstitions' in some sections of the people, that made the return to Hellenic traditions look dangerous. That is the way creative disjunctions between old and new cosmologies usually come about and that is the way such changes are usually justified. Attempts by conventional, 'normal' scientists or historians to paint such returns to the past as fundamentalist or retrogressive cannot stop such time-travel. Returns to the European past of course are more equal than returns to non-European pasts, and I doubt if this argument will convince either the historians of science or the students of fundamentalism.

In the century that is now coming to an end, on a very conservative estimate, more than 170 million people have died of man-made violence. A huge majority of these victims have gone to their graves listening to the paeans being sung by their killers to ideas such as scientific history, development, social evolution, public hygiene and eugenics. Neither European colonialism nor Nazi racism nor Stalinist terror lacked legitimation in scientific rationality and theories of progress. Even the highly skilled, dedicated perpetrators of the last great Third World genocide, the one that took place in Cambodia in the 1970s, showed First World expertise in justifying themselves in the language of scientised history. The Jewish experience with biology in the Third Reich, the Japanese experience with the idea of nuclear experimentation at Hiroshima, and the Russian and the Cambodian peasants' experience with scientific development under scientific socialism have been no different from the experience of the 63 million animals that are sacrificed in the scientific laboratories of the United States every year.[3]

Indeed, one sometimes suspects that the use of terms such as 'colonialism' and 'racism', however appropriate in other ways, has also become an easy means of hiding the complicity of modern science and the idea of progress in the Satanism of our times. One suspects that their use has become an effective means of protecting the Enlightenment vision and the modern West's world-image from criticisms from outside. Though both modern colonialism and racism have shown a highly developed capacity to find legitimacy in modern science, anti-imperial and anti-racist postures have also served as strategies for throttling the voice of those subjected to the new civilising missions of our times. The imperialists and the racists are safely out there being nasty to their victims. They make it unnecessary for others selflessly fighting for the needy and the victimised to look within to see how domination and dispassionate violence work through their favourite categories.

As must have become obvious, in this game of legitimation and cross-legitimation, the central role has been played by modern scientific rationality, absolutised as the ultimate measure of human freedom and the only pathway to truth. From the beginning, scientific rationality of the Baconian variety was placed outside and above the democratic process. Democracy was valued but also feared as piteously vulnerable to self-opinionated demagogues and unthinking mobs. Scientific rationality, on the other hand, was seen as human reason in its purest form and had, therefore, to have priority over democratic aspirations.[4] Today the reign of experts, the tyranny of revolutionary vanguards, and the despotism of the development regimes and other sundry carriers of professionalised, impersonal knowledge have at heart the principle of rationality that modern science enshrines.[5]

Many believe that this complicity of modern science in the contemporary conspiracies against the poor and the weak is accidental. They reject textual, as opposed to contextual, criticisms of modern science as a form of romantic multiculturalism or facile Third-Worldism. Modern science is a value-neutral tool, the clichéd argument goes, that is at the moment controlled by the powerful and the wealthy, but can be used by the oppressed against their tormentors in the long run. Guided by this faith, many activists in the South, bearing manfully the new brown or black man's burden, have been trying to educate the uneducable in modern science, to free them from their age-old superstitions and to empower them.[6]

Well, the long run has turned out to be very long indeed. A minority is now convinced that the record of modern science can no longer be explained away as an aberration brought about by contextual factors such as the control over modern science exercised by the modern state, Western capitalism, colonialism, or the military–industrial complex. They are now convinced that there is something inherently wrong with the text of this kind of a knowledge system. They feel that to tell the South patiently to wait another four hundred years for modern science to redeem its humane potentialities might be considered a trifle inconsiderate by some of the victims.

Not everyone is a futurist, but all citizens have to be a bit of a futurist to respond to the world around them meaningfully, not only for their own sake but also for the sake of their children and grandchildren. The Southern societies are modernising rapidly and some of them in East Asia are now even setting the pace of the present process of globalisation. (So much so that even their present economic crisis is seen by their neighbours as a disease befitting the majesty of the wealthy and the successful. After all, the crisis seems august enough to threaten global economic order.) But it is doubtful if the citizens of some of the oldest civilisations of the world will take, like fish to water, to the famous proposition of John Maynard Keynes that in the long run we are all dead. In the long run, we are all alive—in our children, families, communities and cultures.

It is the responsibility of the citizen–futurist in that context to defy and subvert the 'inevitable' in the future, which is only another name for a tomorrow that dare not be anything other than a linear projection of yesterday. Students of the future owe it to themselves to create a gap between those whose ideas of the future are modelled on the Wall Street stock exchange or on 19th-century Europe's constipated idea of One World and the more marginalised ideas of the future that could be called contemporary versions or reincarnations of the prophetic. The prophets summoned us not because they had elements of the astrologer in them, but because they dared to dream those dreams that were latent in the rest of us. So when we heard their voice, we also heard an inner voice that they had come to represent. The prophet's genius lay in being able to pierce through out structure of defences, everyday sanity and the tendency to read our past into our future.

Prophecies never come true, but prophets are rarely disowned on that count. Like their distant cousins, the astrologers, who by their predictions increase our mastery more over ourselves than over the future, prophets survive not on the truth of their prophecies, but on the fact that they prophesy. People re-read their texts, adjust their meanings and try to go on as if nothing has changed when a prophecy goes wrong. That is because the prophet's authenticity is not dependent on his or her prophecies. The prophet is a prophet by virtue of creating a gap between our past and present on the one hand, and our futures on the other. The gap keeps alive our scepticism towards the given 'truths' and the 'axioms' of our world, for these are the truths and axioms that have to be defied or transcended, to create the vision of a disjunctive future, a future that refuses to be a linear projection of the past or the present.[7]

This exercise is particularly relevant in Southern societies, whose futures have been hijacked and are being held captive by the developed societies of the West.[8] The West's today is the Third World's tomorrow, exactly as the Third World's today is West's yesterday. Territorial and cultural differences have been flattened into historical stages and no escape route has been kept open for recalcitrant societies trying to avoid being straitjacketed. Conformity in this respect has become the ultimate measure of sanity. Futurism should ideally allow one to break the mould of conventional rationality; it should allow one to claim that exactly as there are various forms of irrationality, there are many forms of rationality waiting to be rediscovered by us.

Notes

1. The present age is global, indeed all-too-global, says Lewis Gordon, but in a tragic way. Lewis Gordon, 'The Tragic Dimensions of our Neocolonial "Postcolonial" world', in Emmanuel Chukwudi Eze (ed.), *Postcolonial African Philosophy: A Critical Reader*, Oxford: Blackwell, 1997, pp. 241–51 (p. 243).
2. E.g. Deviprasad Chattopadhyay, *Lokayata: Study in Ancient Indian Materialism*, New Delhi: Peoples Publishing House, 1976; Amartya Sen, 'Indian Tradition and the Western Imagination', *Daedalus*, 126(2): 1–26, spring 1997. See on this issue Ashis Nandy, *The Intimate Enemy: Loss and Recovery of Self Under Colonialism*, New Delhi: Oxford University Press, 1983, ch. 1.
3. Shiv Visvanathan, 'On the Annals of the Laboratory State', in Ashis Nandy (ed.), *Science, Hegemony and Violence: A Requiem for Modernity*, Tokyo and New Delhi: United Nations University and Oxford University Press, 1988, pp. 257–88 (see p. 266).
4. E.g. the essays in Ziauddin Sardar (Ed.), *The Revenge of Athena: Science, Exploitation and the Third World*, London: Mansell, 1988; and Nandy, *Science, Hegemony and Violence*. Also Vandana Shiva, *Staying Alive: Women, Ecology and Development*, New Delhi: Kali for Women, 1989; Claude Alvares, *Science, Development and Violence*, New Delhi: Oxford University Press, 1992.
5. Some of the best analyses of the problem are in the works of Ivan Illich. See e.g. *Disabling Professions*, London: Marion Boyars, 1900. Also Graham Hancock, *Lords of Poverty*, London: Mandarin, 1989.
6. A series of non-governmental organisations have come up in South Asia during the last twenty years to propagate the scientific temper or spirit among the recalcitrant citizens of the various countries in the region. The KSSP of Kerala has been the largest and probably most successful of them.
7. Ashis Nandy, 'Shamans, Savages and the Wilderness: On the Audibility of Dissent and the Future of Civilisations', *Alternatives*, 14(3): 263–77, July 1989.
8. Ziauddin Sardar, 'Colonising the Future: The "Other" Dimension of Future Studies', *Futures*, 25(2): 179–87, March 1993.

18
Other Futures Studies: A Bibliographic Essay

MERRYL WYN DAVIES

There is an overabundance of literature on the future as well as futures studies. But, as contributors to this book have argued, most of this literature is Eurocentric. It sees the future as a Western challenge and a Western opportunity: the future will be shaped, is being shaped, essentially in the image of the West. This monolithic projection of the future is most common among airport 'bestsellers' that originate from management stables, such as *Rethinking the Future*,[1] and the massive techno-junkie output, such as *Wired* magazine's *Reality Check*.[2] For the former, 'rethinking' actually means rethinking competition, markets, management and leadership in an attempt to maintain the global status quo; for the latter, 'real future' is about 'movies on demand', 'male birth control pills', the 'orgasmatron', 'robot surgeons', 'smart fabrics' and 'self-cleaning toilets'. These books vie for attention with works by more conventional futurists, the likes of Alvin Toffler and John Naisbitt. Here the essential game is prediction: Naisbitt's *Megatrands* and *Megatrends Asia* (most of the predictions here are already in trouble!)[3] and Barry Howard Minkin's *Future in Sight*[4] reduce futures studies to a string of self-fulfilling prophecies—hardly an improvement on *Nostradamus: The Next 50 Years*[5] or *The Millennium Planner*,[6] which glorifies Nostradamus, Malachy, Edgar Cayce and Jean Dixon!

Sardar has criticised the Eurocentrism, in the form of the Pax Americana, and technology fetish that so dominates the output of the Washington-based World Future Society.[7] Its study manuals, guides to futures studies and books based on its annual conferences seem to show no awareness of the world outside America. However, this is just as true of many futures studies manual coming from Europe. To give a few examples: the 'Next Three Futures' contemplated by Warren Wagar seem to leave the non-West out in the cold;[8] *Into the 21st Century*,[9] subtitled 'A Handbook for a Sustainable Future', by Brian Burrows *et al.* is little more than a compendium of the output of the environmental movement in Europe and America; even Bruce Lloyd's excellent anthology of papers from *Futures* could have been more balanced towards the non-West.[10]

But of course not all futures literature emanating from Europe and America is myopic and shallow. Hazel Henderson's *Building a Win–Win World*,[11] while concerned largely with America, has a strong multicultural flavour; a particular feature of Henderson's work, apart from her Herculean efforts to undermine mainstream economics, is her concern for the developing world. Almost all her books suggest ways we can rethink development; in this respect, *Paradigms in Progress*,[12] is particularly good. There are also dissenting views of the future within the Western paradigm. Robert Heilbroner's *Visions of the Future*[13] and the late Jean Gimpel's *The End of the Future*[14] offer a refreshing break from the dominant technology-obsessed projections of Western culture on the future while attempting to humanise our thoughts about tomorrow. And for a really good Western critique of Western notions of futures, Barry Hughes's *World Futures: A Critical Analysis of Alternatives*[15] would be hard to beat. Hughes shows how the worldview of American futurists is deeply ingrained in their theoretical assumptions and the selection and interpretation of their data, so that their predictions in fact are little more than a reflection of their prejudices. In other words, short-sightedness feeds on itself.

Pluralistic Works

Fortunately, the efforts of many non-Western futurists are now beginning to bear fruit and pluralistic and multicultural futures literature is steadily growing. We can see this in what is perhaps considered to be the foundational work in this field: *The Knowledge Base of Futures Studies.*[16] Richard Slaughter and his collaborators have set out deliberately to open up the futures field to multicultural and pluralistic concerns. The result is truly joyful: here one finds powerful articulations of non-Western ideas, notions and perspectives on the future. The multi-volume project has its origins in the special issue of *Futures* that was intended to develop a core understanding of futures studies. The first three volumes concentrate on 'Foundations', 'Organisations, Practices, Products' and 'Directions and Outlooks'. Slaughter argues that knowledge base of futures studies is an evolving process;[17] and later volumes will expand the field and bring a great deal of fresh material from non-Western and dissenting perspectives. In contrast, Wendell Bell's two-volume study *Foundations of Futures Studies*[18] is much too American, although the second volume does attempt to open the field to ethical questions and the thorny issues of good society. These foundational studies are well complemented by Eleonora Masini's neat introductory text *Why Futures Studies?*[19] Masini incorporates the scholarship of many non-Western futurists—including Ashis Nandy, Ziauddin Sardar, Sohail Inayatullah and Susantha Goonatilake—into her analysis and discussion of futures methods, principles and concepts.

There are also some good undergraduate futures texts that show awareness of other cultures, and some that are unconsciously pluralistic if not overtly so. *Futures: Concepts and Powerful Ideas*[20] by Richard Slaughter has gone through several mutations, revisions and critical poundings. Slaughter uses the techniques of cultural studies to take his readers through the entire field of futures. Frank Hutchinson's *Educating beyond Violent Futures*[21] tries to do just what the title suggests. Hutchinson combines peace studies and educational theory to give a certain edge to futures studies. Both these books are well illustrated and use charts and diagrams to communicate complex ideas well. Graham May's *The Future Is Ours*[22] is aimed at more advanced students. May tries to combine foresight, management and creating futures into a viable synthesis and relies heavily on case studies. May is not very critical but is aware of epistemological debates and issues. Allen Tough's *Crucial Questions about the Future*[23] teaches by asking leading questions. Tough wants readers to question their own assumptions, biases and perceptions about the future and then leads them through different pathways to alternative futures. More politically overt futures can be found in 'Back to the Future', a special issue of the left-wing campaigning magazine *New Internationalist*.[24] It provides an interesting blend of writers, artists and poets reflecting on the future with well-illustrated graphics and data boxes, all with strong emphasis on cultural pluralism and social justice. 'Holistic Education: Preparing for the 21st Century', a special issue of the quarterly magazine *New Renaissance*,[25] does a similar job from a spiritual and transpersonal perspective. Paul Wildman and Sohail Inayatullah try to make sense of these different approaches to teaching the future in their fascinating paper 'Ways of Knowing, Civilization, Communication and the Pedagogies of the Future'.[26] They examine case studies of futures teaching in a multicultural setting and discuss the different ways human beings learn about their world and the future. The authors argue that futures studies should be taught in an open and multi-layered way.

Visions of people who would normally be considered the 'periphery' have now begun to move towards 'the centre'. In 'What Futurists Think', a special issue of *Futures*,[27] Sohail Inayatullah brings a veritably multicultural collection of futurists together. Fifty futurists, from virtually every part of the world, describe their visions and reveal why they became interested in futures studies in the first place. This anthology contains some of the most beautiful and subtle writing on the future that one is likely to find anywhere. Beauty in profoundly simple form can also be found in *Vision Aotearoa*,[28] which presents, in conversation, visions of the future of New Zealand (Aotearoa) from Maori and Pakeha perspectives. How visions can take you beyond the ordinary is also highlighted in *Voices of the First Day*,[29] an inspiring look at Aboriginal Dreamtime and the Aboriginal way of knowing. *In Seeds of Peace*,[30] Sulak Sivaraska presents 'a Buddhist vision for renewing society' and creates a vision of alternative futures based on self-reliance, humanistic concern for the Other, social justice and a variety of concepts that emerge from the eightfold path of Buddha. Sivaraska insists that capitalism and Buddhism cannot exist together and charts a strategy through which we can all transform our inner and outer selves.

One could consider these visions to be those of the periphery. But, of course, the 'periphery' does not always exist outside 'the centre'; 'the centre' can also have its own periphery. The marginalised in the metropolitan centres of Europe are like all marginalised people everywhere—silenced and shut out from opportunities. *Visions of Europe*[31] tries to re-examine the legacy of Europe, in particular its tendency to marginalise certain classes and demonise outsiders, and articulates its futures with a strong accent on multiculturalism. In the book, Richard Kearney, Irish philosopher and broadcaster, examines viable European futures in conversation with a collection of noted writers and thinkers, including Vaclav Havel, Edward Said and Seamus Heaney. The end product gives you hope that Europe can transcend itself despite its legacy. And so can the Western world itself—if the Millennium Project's more recent efforts are to be believed. Over 200 thinkers from all over the world contributed to *The State of the Future*.[32] The report presents global thinking on eighteen issues, which are developed through cross-impact analysis and scenario development. Dissenters are respected; and their positions are not ignored but given ample space. Previously, all such exercises were highly Eurocentric; but here there is a genuine effort to move out of the narrow confines of the West. The result is a positive step towards developing a collective global vision and a global early warning system.

Asian and Islamic Perspectives

There is no dearth of good scholarship on futures from the Asian perspective. To begin with there are two special issues of *Futures* specifically devoted to South Asian futures. 'The Futures of South Asia',[33] edited by Sohail Inayatullah, explores images of the region's futures and the problems with creating futures in an area dominated by history and nationalist politics. 'South Asia: Fifty Years on',[34] edited by Ziauddin Sardar, looks at representation of the Sub-continent, and explores how the Sub-continent can extricate itself from the quagmire of fragmentation, nation-state politics and ethnic rivalries. Both include essays on Pakistan, India, Bangladesh and Sri Lanka. An excellent exploration of Asian futures is provided by Yogesh Atal and Eleonora Masini's anthology *The Futures of Asian Cultures*,[35] based on a lively Unesco-sponsored seminar in Bangkok. The book is marred by poor editing and typographical errors but it provides an unparalleled assessment of the sources of cultural vitality in Asia. Asian perspectives are combined with African and Latin American, together with American and European, in *The Futures of Cultures*.[36] Masini toiled for almost a decade, organising seminars on regional cultural futures with the assistance of Unesco. This volume distils the best of these seminars. New thinking on futures from Malaysia, Singapore, Indonesia and South Korea is brought together in *Asia in the 21st Century*[37]—the result of a study by the Malaysian Strategic Research Centre. Malaysia has been very active in the futures field. And the ideas and works of both Prime Minister Mahathir Mohamad and Deputy Minister Anwar Ibrahim have had considerable influence on futures thinking in Southeast Asia. Their visions and approach to the future could not be more different. Mahathir's outlook is rather technological and based on an outspoken anti-West stance, as is so evident in *The Voice of Asia: Two Leaders Discuss the Coming Century*.[38] In dialogue with Shintaro Ishihara, the author of *The Japan That Can Say No*,[39] Mahathir discusses the 'Pacific Century', Western hedonism and the new Asian paradigms. Anwar's vision is grounded within Asian traditions and emphasises 'symbiosis'. *The Asian Renaissance*[40] paints a picture of the future of Asia based on civil society, humane ethics, symbiosis between East and West and primacy of indigenous cultures.

An appreciation of Ashis Nandy's thought is essential for understanding Indian futures. Ziauddin Sardar's essay 'The A, B, C, D (and E) of Ashis Nandy'[41] is an excellent starting-point for appreciating the notion of alternatives and dissent in Nandy's thought. A special issue of *Emergences*, 'Plural Worlds, Multiple Selves: Ashis Nandy and Post-Colombian Future',[42] contains a host of papers offering reasoned analysis of Nandy's thought from a variety of perspectives, as well as some of his own more neglected essays. However, there is no substitute for reading Nandy himself. *The Intimate Enemy*[43] analyses how deeply colonialism has effected and infiltrated the Indian psyche—any viable future for India must confront the loss of Self under colonialism. *Traditions, Tyranny and Utopias*[44] shows how yesterday's utopias become today's nightmares and discusses possible alternatives to the instrumental rationality of modern science and technology. A concise account of Nandy's philosophy of the future can be found in 'Bearing Witness to the Future';[45] a more elaborate discussion of the same ideas is available in 'Shaman, Savages and the Wilderness: On the Audibility of Dissent and the Future of Civilisations'.[46]

When it comes to Islam, the work of Ziauddin Sardar is indispensable. Through a series of books and papers, Sardar has explored Islamic futures from a number of different perspectives; indeed, it seems that he has made the field his own. His classic studies *The Future of Muslim Civilisation*[47] and *Islamic Futures*[48], both of which have been translated into several languages, present his own radical philosophy on Islam and futures studies. *The Future of Muslim Civilisation* offers a grand vision of Muslim civilisation based on the original egalitarian thought of Islam before it was undermined in the aftermath of Islam's political success. Sardar argues that Islam has to be rediscovered, indeed reinvented, just as traditions reinvent and renovate themselves, in every epoch of human history. This rediscovery takes place within the conceptual parameters of Islam—so Muslim societies change and yet remain themselves. When first published, this study was years ahead of its time. Only now is its true impact beginning to be felt in Muslim societies. *Islamic Futures* takes a critical look at a host of new ideas that have emerged over the last few decades—from the notion of 'Islamic state' to the ideas of 'Islamic economics' and 'Islamic science' to new development strategies to the global discourse of 'Islamicisation of knowledge'. Sardar puts these ideas under a futures microscope, reveals their shortcomings, suggests alternatives and argues that creative futures that are more holistic and ecologically sensitive can emerge from Islam. In *An Early Crescent*,[49] edited by Sardar and subtitled 'The Future of Knowledge and the Environment in Islam', many of these new ideas and discourses are discussed by their main proponents and cast within the framework of futures. This volume is based on a major international Islamic conference entitled 'Dawa and Development in the Muslim World: The Future Perspective', which was held in Makkah, Saudi Arabia, in 1987. The conference, organised by Sardar with the sponsorship of the Muslim World League, also produced two sister volumes. *Today's Problems, Tomorrow's Solution*,[50] edited by Abdullah Naseef, explores future structures of Muslim societies, and *Beyond Frontiers*,[51] edited by Merryl Wyn Davies and Adnan Pasha, examines Islam and contemporary needs. The special issue of *Futures* 'Islam and the Future',[52] also edited by Sardar, contains a more recent examination of how the alternative futures universe of Islam is shaping up. A study that is essential to understanding the contemporary and future issues of ethics, gender and technology in Muslim societies is Munawar Anees's *Islam and Biological Futures*.[53] Anees looks at how Muslim societies coped with such issues in history and gives pointers to how these problems can be solved in the future.

Other Dissenting Views

Africa has not been so well represented in futures studies. But the situation is definitely improving. In understanding African futures, it is worthwhile to start with Godwin Sogolo's *Foundations of African Philosophy*.[54] Sogolo is a futurist and looks at African philosophy from the futures perspective. Odera Oruku's *Sage Philosophy: Indigenous Thinkers and the Modern Debate on African Philosophy*[55] shows how African philosophy is intricately linked with African futures. He interviews a number of African sages, analyses their thoughts on the notions of self, community and future, and locates the emerging discourses in relation to the relative influences of Islam and the West on African traditions and philosophy. Oruku argues that the future of Africa has to be shaped by its multiple traditions. What would African futures based on these multiple traditions be like? The answers can be found in *Visions of the Future of Africa*,[56] edited by Olugbenga Adesida and Arunma Oteh. More specific visions of African futures, including a few prescriptive ones, are contained in the papers included in the proceedings of the Fourteenth World Conference of the World Futures Studies Federation, held in Nairobi in 1995: *Futures beyond Poverty*.[57]

Like Africa, feminism too has been slow in making an impact on futures studies. It is indeed surprising that serious attempts have not been made to introduce the notion of the feminine into futures studies; although there has been a strong feminist presence in the futures network. Scholars such as Magda McHale, Eleonora Masini, Hazel Henderson and Katrin Gillward have had a tremendous influence on the international futures movement, particularly the World Futures Studies Federation. However, only two noteworthy attempts to develop a specific feminist epistemology for futures research have been made so far—by Vuokko Jarva[58] and Ivana Milojevic.[59] Jarva argues that female futures research is focused not so much on problem-solving but on enabling women to realise their own desired futures; Milojevic suggests that feminist futures should be empowering; both argue that futures work should be gender conscious. In some cases the feminine enters the futures field through spirituality. For example, Tadbhavananda Avadhuta and Jayanta Kumar's[60] examination of Indian history through the religious figures of Shiva and Krishna produces a rather new and feminine vision of future—a vision that most feminists would appreciate.

When looking at technological trends, it is always worth looking at dissenting views. For it is almost always the dissenters, whether from the West or the non-West, who speak for the marginalised, temper the hyperbole and bring the reality of the non-West into the debate. An antidote to all the hype about cyberspace, for example, is provided by *Cyberfutures*,[61] which explores 'culture and politics on the information superhighway'. Contributors to this anthology discuss democracy, religion, nature and consciousness and argue that cyberspace is little more than a new wave of Westernisation. In an essay of formidable power entitled 'alt.civilisation.faq—Cyberspace as the Darker Side of the West', Ziauddin Sardar shows how the colonisation of cyberspace is actually repeating the imperial patterns of the age of Empire. Other contributors suggest that the future will be bifurcated into two varieties of community, one experiencing more intoxicating powers, the other deeper and deeper hopelessness. Many of the ideas in *Cyberfutures* are echoed in *Cultural Ecology: The Changing Dynamics of Communications*.[62] Danielle Cliche's anthology offers a brilliant deconstruction of the futuristic claims of Bill Gates, shows how new technologies continue the hegemony of multinational corporations by reducing civil and national liberties and undermining democracy, and argues that a global conversation of cultures that goes beyond the hype of the information society is needed to bring sanity into the so-called information revolution. The vision and challenge of the future cannot be the creation of an information society but shaping a genuine cultural pluralism based on appreciation and participation of all cultures of the world. I cite these books only as an example that alternatives to dominant visions do exist—both in the West and the non-West.

Non-Western futures are intrinsically linked to the recovery of non-Western history. A couple of futuristic histories have been produced recently with this very aim. In *Macrohistory and Macrohistorians*,[63] Johan Galtung and Sohail Inayatullah attempt to produce a general theory of macrohistory and try to show how macrohistory affects the future. They offer a comparative analysis of twenty macrohistorians—including ibn Khaldun, Comte, Vico, Marx, Hegel, Ssu-Ma Chien, Sarkar, Toynbee, Weber and Sorokin—in order to derive futures lessons from history. In contrast, Ziauddin Sardar, Merryl Wyn Davies and Ashis Nandy argue that the last five hundred years of colonial history have trapped us into moving in ever-decreasing, ever-constricting, concentric circles. Decolonising the future of the non-West as well as liberating the West from itself requires us to confront the suffocating embrace of this history. The challenge, concludes *Barbaric Others*,[64] is to 'recover our plural pasts, and, through them, our plural futures so that once again there can begin to be histories, metahistories and mythographies as various and equally valid ways of seeing the world and constructing the past, and as equally valid responses to events and ideas that sustain a plural future for the world's people in all their diversity'.

A great deal of non-Western futures scholarship is an outcome of the critique of the West. Much of Ashis Nandy's work, for example, is in this vein. Ziauddin Sardar has also used his critique of Western ideas and disciplines to suggest alternative paths to the future; for example, in his series of brilliant articles on the films of Arnold Schwarzenegger,[65] his survey of non-Western critique of human rights and democracy,[66] and his deconstruction of development[67] and the notion of discipline in the Western framework of knowledge.[68] Similarly, Tae-Chang Kim[69] has used his criticism of the West and all things Western, including rationalism, logocentrism and technology fetish, to generate a whole new variety of futures discourse—the study of future generations.[70] Kim argues for 'a global family person'; and his specifically Korean perspective, based so deeply in holism, leads him to suggest that we should view ourselves 'as members of the global family' in which the 'past, present, and future generations are also family members of our home, Earth'. The study of future generations is now threatening to become a whole new field of future studies.[71]

Other Time, Other Canons

A genuine understanding of non-Western attitudes towards the future must incorporate Other notions of time. In Western civilisation time is essentially linear: past, present, future. The future can even mark a sharp dislocation from the past and the present; the present may not always incorporate the past. But for non-Western societies, the past is ever present as dynamic, life-enhancing tradition. Moreover, non-Western cultures insist on adjusting to change, moving to the future, with their identities and traditions intact—that is, changing but remaining themselves. This necessitates a new understanding of the future: the future becomes an amalgam of the past and the present and is not always linear; it may be cyclic, weaved as a tapestry or connected to the soul. Thus, in many non-Western societies, time is embedded in social interactions, cultural practices and structures of knowledge, as well as the environment. There are no specific studies of non-Western time in futures studies—but Sohail Inayatullah's essay 'Framing the Shapes and Times of the Future'[72] makes an excellent first attempt at bringing other notions of time within the framework of futures studies. For a wider appreciation of time, Jacob Needleman's brief book *Time and the Soul*[73] is indispensable. Needleman focuses on the spiritual dimension of time and compares the notion of time in Greek, Indian and other philosophies, and he succeeds in showing that if we understood what is eternal about life and work we could change our relationship to time as well as our perception of it. All that anyone would even want to know on *Time in Indian Philosophy*[74] can be found in Hari Shankar Prasad's truly massive anthology. However, for most people, the opening essay, which provides a summary of the book, should be sufficient. Anthony Aveni, who is particularly impressive on the notion of time in pre-Columbian and tribal societies, provides a good discussion of linear and cyclic time, and the relationship between time-keeping and political power, in *Empires of Time.*[75] Mircea Eliade produced a classic in *The Myth of the Eternal Return,*[76] which examines time from a number of different religious traditions, revealing the underlying structure of the spiritual world-view. A number of recent studies, while remaining within the Western framework, do manage to show that time need not be linear. Barbara Adam's *Timewatch*[77] focuses on the social analysis of time, bringing in areas of health, education, work and environment; and 'Time and Space: Geographic Perspective on the Future',[78] a special issue of *Futures* edited by Michael Batty and Sam Cole, attempts to integrate all those disciplines that 'share time-mapped lives' to bring an inter-disciplinary, time-based perspective on future studies.

Apart from the notion of time, those concerned with pluralistic and non-Western futures would do well to familiarise themselves with some of the major works of non-Western civilisations, including a few of their more noted thinkers and writers. Here, I would suggest only two works each from Islamic, Chinese, Indian and Japanese civilisations; excluding the religious texts such as the Qur'an and the Mahabharata, which anyone interested in other cultures must surely peruse.

Ibn Khaldun's *The Muqaddimah,*[79] more commonly known as *Introduction to History*, is essential reading for anyone interested in rise and fall of civilisation—particularly from the Islamic viewpoint. Ibn Khaldun is, of course, the founder of sociology, who had serious influence on Comte, Weber, Marx, Toynbee and numerous other historians and sociologists. Al-Ghazzali's *The Book of Knowledge*[80] contains one of the earliest discussions of science and values and provides a deep insight into Muslim intellectual consciousness and the long historic debate about sources of knowledge. *The Analects* of Confucius[81] is probably the most essential text on Chinese civilisation, followed closely by *The Book of Menicus.*[82] Confucius insists that good influence is of greater value in politics than force and that social relationships should be governed by natural ties and emphasis on social harmony. Menicus focuses on righteousness, love, justice and fairness and argues, most sensibly, that ordinary people are much more important than kings and princes; he had a deep sense of what is a good society and great respect for ordinary folks—an ancient futurist if there ever was one! Kalidasa's drama *Sakuntala*[83] is a masterpiece of Indian literature. The first part of this complex work, the lyrical meditation on nature 'The cloud messenger', would be of particular interest to futurists. Two recent Indian writers should be read as one; both have shaped the modern Indian consciousness and are indispensable for understanding Indian futures. Rabindranath Tagore's various poems and plays[84] (his novels are not as good) and Mohandas Gandhi's *Autobiography*[85] present two contrasting views of the Indian tradition in its encounter with the West; the future of India cannot escape a synthesis of their ideas. *The Pillow Book*[86] by Sei Shonagon gives invaluable observations on Japanese thought and aesthetic sensibilities; and Yoshida no Kenko's *Essays on Idleness*[87] extends these observations to nature, art, society and the lifestyle of monks.

All of these works provide fundamental insights into the ways of knowing, being and doing of non-Western civilisations—without some appreciation of non-Western canons, I contend, it is difficult to do futures work across cultures and civilisations. But, of course, this is only the tip of the iceberg. More detail on non-Western classics and non-Western thinkers can be found in *A Guide to Eastern Literatures*,[88] edited by David Marshall Lang; *Great Thinkers of the Eastern World*,[89] edited by Ian McGreal; and the *Encyclopaedia of the History of Science, Technology, and Medicine in Non-Western Cultures*,[90] edited by Helaine Selin. *The Dictionary of Global Cultures*,[91] edited by Kwame Anthony Appiah and Henry Louis Gates, Jr. is a handy reference work to have around, even though it has many omissions, errors and idiosyncrasies.

Three writers and scholars have made an inaudible mark both on opening up futures studies to pluralism and furnishing us with relevant perspectives on non-Western futures: Ziauddin Sardar, Ashis Nandy, Sohail Inayatullah. It is worth noting that none of these futurists made it to the 'One hundred must influential futurists' listed in *Encyclopaedia of the Future*[92]—the two-volume, grotesquely mindless celebration of the Pax Americana. Clearly some futurists live on another world!

Notes

1. Rowan Gibson, *Rethinking the Future*, London: Nicholas Brealey, 1997.
2. Brad Wieners and David Pescovitz, *Reality Check*, San Francisco, CA: Hardwired, 1996.
3. John Naisbitt, *Megatrends*, New York: Warner Books, 1984; John Naisbitt, *Megatrends Asia: The Eight Asian Megatrends That Are Changing the World*, London: Nicholas Brealey, 1996.
4. Barry Howard Minkin, *Future in Sight: 100 of the Most Important Trends, Implications and Predictions for the New Millennium*, New York: Macmillan, 1995
5. Peter Lemesurier, *Nostradamus: The Next Fifty Years*, London: Peter Lemesurier, 1993.
6. Peter Lorie, *The Millennium Planner*, London: Boxtree, 1995.
7. Ziauddin Sardar, 'Colonising the Future: The "Other" Dimension of Future Studies', *Futures*, 25(3), 1993.
8. W. Warren Wagar, *Next Three Futures: Paradigms of Things to Come*, London: Adamantine Press, 1992.
9. Brian Burrows, Alan Mayne and Paul Newbury, *Into the 21st Century: A Handbook for a Sustainable Future*, London: Adamantine Press, 1991
10. Bruce Lloyd (ed.), *Framing the Future: A 21st Century Reader*, London: Adamantine Press, 1998.
11. Hazel Henderson, *Building a Win–Win World: Life beyond Global Economic Warfare*, San Francisco, CA: Berrett-Koehler, 1996.

12. Hazel Henderson, *Paradigms in Progress: Life Beyond Economics*, Indianapolis: Knowledge Systems Inc., 1991.
13. Robert Heilbroner, *Visions of the Future*, New York: Oxford University Press and New York Public Library, 1995.
14. Jean Gimpel, *The End of the Future: The Waning of the High-Tech World*, London: Adamantine Press, 1995.
15. Barry B. Hughes, *World Futures: A Critical Analysis of Alternatives*, Baltimore, MD: Johns Hopkins University Press, 1985.
16. Richard Slaughter (ed.), *The Knowledge Base of Futures Studies*, 3 vols. (more planned), Victoria: DDM Media Group, 1996.
17. Richard Slaughter, 'The Knowledge Base of Future Studies as an Evolving Process', *Futures*, 28(9): 799–812, November 1996.
18. Wendell Bell, *Foundations of Futures Studies*, 2 vols., New Brunswick, NJ: Transaction, 1997.
19. Eleonora Barbieri Masini, *Why Futures Studies?*, London: Grey Seal, 1993.
20. Richard A. Slaughter, *Futures: Concepts and Powerful Ideas*, 2nd edn., Victoria: Futures Study Centre, 1996 (1st edn. 1991).
21. Frank Hutchinson, *Educating Beyond Violent Futures*, London: Routledge, 1996.
22. Graham H. May, *The Future Is Ours*, London: Adamantine Press, 1996.
23. Allen Tough, *Crucial Questions about the Future*, Lanham, MD: University Press of America, 1991.
24. 'Back to the Future', special issue of *New Internationalist*, 269, July 1995.
25. 'Holistic Education: Preparing for the 21st Century', special issue of *New Renaissance*, 6(3), 1996.
26. Paul Wildman and Sohail Inayatullah, 'Ways of Knowing, Civilization, Communication and the Pedagogies of the Future', *Futures*, 28(8): 723–40, October 1996.
27. Sohail Inayatullah (ed.), 'What Futurists Think', special issue of *Futures*, 28(6–7), July–August 1996.
28. Roslie Capper, Amy Brown and Witi Ihimaera (in conversation), *Vision Aotearoa*, Wellington: Bridget William Books, 1994.
29. Robert Lawlor, *Voices of the First Day: Awakening in the Aboriginal Dreamtime*, Vermont: Inner Traditions, 1991.
30. Sulak Sivaraska, *Seeds of Peace: A Buddhist Vision for Renewing Society*, Berkeley, CA: Parallax Press, 1992.
31. Richard Kearney, *Visions of Europe*, Dublin: Wolfhound Press, 1992.
32. Jerry Glenn and Theodore Gordon (eds.), *The State of the Future: Implications for Actions Today*, Washington, DC: American Council for the United Nations University, 1997.
33. Sohail Inayatullah (ed.), 'The Futures of South Asia', special issue of *Futures*, 24(9), November 1992.
34. Ziauddin Sardar (ed.), 'South Asia: Fifty Years on', special issue of *Futures*, 29(10), December 1997.
35. Eleonora Masini and Yogesh Atal (eds.), *The Futures of Asian Cultures*, Bangkok: Unesco, 1993.
36. Eleonora Masini and Albert Sasson (eds.), *The Futures of Cultures*, Paris: Unesco, 1994.
37. Abdul Razak Abdullah Baginda (ed.), *Asia in the 21st Century*, Selangor: Pelanduk Publications (M) Sdn. Bhd., 1996.

38. Mahathir Mohamad and Shintaro Ishihara, *The Voice of Asia: Two Leaders Discuss the Coming Century*, Japan: Kodansha International Ltd., 1995.
39. Shintaro Ishihara, *The Japan That Can Say No*, New York: Simon & Schuster, 1989.
40. Anwar Ibrahim, *The Asian Renaissance*, Singapore: Times Books International, 1996.
41. Ziauddin Sardar, 'The A, B, C, D (and E) of Ashis Nandy', *Emergence*, 7–8:126–45, 1995–6; also *Futures*, 29(7): 649–60, September 1997.
42. Vinay Lal (ed.), 'Plural Worlds, Multiple Selves: Ashis Nandy and Post-Colombian Future', *Emergence*, 7–8, 1995–6.
43. Ashis Nandy, *The Intimate Enemy: Loss and Recovery of Self under Colonialism*, Delhi: Oxford University Press, 1983.
44. Ashis Nandy, *Traditions, Tyranny and Utopias*, Delhi: Oxford University Press, 1992.
45. Ashis Nandy, 'Bearing Witness to the Future', *Futures*, 28(6–7): 636–9, August–September 1996.
46. Ashis Nandy, 'Shaman, Savages and the Wilderness: On the Audibility of Dissent and the Future of Civilisations', *Alternatives*, 14(3), July 1989.
47. Ziauddin Sardar, *The Future of Muslim Civilisation*, London: Croom Helm, 1979; 2nd edn., London: Mansell, 1987.
48. Ziauddin Sardar, *Islamic Futures: The Shape of Ideas to Come*, London: Mansell, 1985.
49. Ziauddin Sardar, *An Early Crescent: The Future of Knowledge and Environment in Islam*, London: Mansell, 1989.
50. Abdullah Naseef (ed.), *Today's Problems, Tomorrow's Solution*, London: Mansell, 1988.
51. Merryl Wyn Davies and Adnan Pasha (eds.), *Beyond Frontiers: Islam and Contemporary Needs*, London: Mansell, 1989.
52. Ziauddin Sardar (ed.), 'Islam and the Future', special issue of *Futures*, 23(3), April 1991.
53. Munawar Anees, *Islam and Biological Futures*, London: Mansell, 1989.
54. Godwin Sogolo, *Foundations of African Philosophy*, Ibadan: Ibadan University Press, 1993.
55. Odera Oruku, *Sage Philosophy: Indigenous Thinkers and the Modern Debate on African Philosophy*, Nairobi: Acts Press, 1991.
56. Olugbenga Adesida and Arunma Oteh (eds.), *Visions of the Future of Africa*, London: Adamantine Press, 1999.
57. Gilbert Ogutu, Penitti Malaska and Johanna Kojola (eds.), *Futures Beyond Poverty: Ways and Means out of the Current Stalemate*, Turku: Finland Futures Research Centre, 1995.
58. Vuokko Jarva, 'Toward Female Futures Research', in Mika Mannermaa, Sohail Inayatullah and Rick Slaughter (eds.), *Coherence and Chaos in Our Uncommon Futures*, Turku: Finland Futures Research Centre, 1994.
59. Ivana Milojevic, 'Towards a Knowledge Base for Feminist Futures Research', in Richard A Slaughter (ed.), *The Knowledge Base of Futures Studies*, vol. 3, Victoria: DDM Media Group, 1996.
60. Tadbhavananda Avadhuta and Jayanta Kumar, *The New Wave*, Calcutta: Proutist Universal, 1985.
61. Ziauddin Sardar and Jerome Ravetz (eds.), *Cyberfutures: Culture and Politics on the Information Superhighway*, London and New York: Pluto Press and New York University Press, 1996.

62. Danielle Cliche, *Cultural Ecology: The Changing Dynamics of Communications*, London: International Institute of Communications, 1997.
63. Johan Galtung and Sohail Inayatullah, *Macrohistory and Macrohistorians*, New York: Praeger, 1997.
64. Ziauddin Sardar, Merryl Wyn Davies and Ashis Nandy, *Barbaric Others: A Manifesto on Western Racism*, London: Pluto Press, 1993.
65. Ziauddin Sardar, 'Surviving the Terminator: The Postmodern Mental Condition', *Futures*, 22(2): 203–10, March 1990; Ziauddin Sardar, 'Terminator 2: Modernity, Postmodernity and Judgement Day', *Futures*, 25(5): 493–506, June 1992; and Ziauddin Sardar, 'Total Recall: Aliens, Others and Amnesia in Postmodernism', *Futures*, 23(2): 189–203, March 1991.
66. Ziauddin Sardar, 'Future Challenges of Human Rights and Democracy', *Futuresco*, 5, Paris: Unesco, 17–24 June 1996.
67. Ziauddin Sardar, 'Beyond Development: An Islamic Perspective', *European Journal of Development Research*, 8(2), 1996.
68. Ziauddin Sardar, 'Development and the Locations of Eurocentrism', in Ronaldo Munck and Denis O'Hearn (eds.), *Critical Holism: (Re)Thinking Development*, London: Zed, 1998.
69. Tae-Chang Kim, 'Coherence and Chaos in Our Uncommon Futures: A Hun-Philosophical Perspective', in Mika Mannermaa, Sohail Inayatullah and Rick Slaughter (eds.), *Coherence and Chaos in Our Uncommon Futures: Visions, Means, Actions*, Turku: World Futures Studies Federation, 1994, pp. 41–8.
70. Tae-Chang Kim, 'Toward a New Theory of Value for the Global Age', in Tae-Chang Kim and James Dator, *Creating a New History for Future Generations*, Kyoto: Institute for the Integrated Study of Future Generations, 1994.
71. Tae-Chang Kim and James A. Dator (eds.), *Co-Creating Public Philosophies for Future Generations*, London: Adamantine Press, 1998; and Richard Slaughter and Allen Tough (eds.), 'Learning and Teaching about Future Generations', special issue of *Futures*, 29(8), October 1997
72. Sohail Inayatullah, 'Framing the Shapes and Times of the Future', in R. Slaughter (ed.), *The Knowledge Base of Futures Studies*, vol. 3, Victoria: DDM Media Group, 1996.
73. Jacob Needleman, *Time and the Soul*, New York: Doubleday, 1998.
74. Hari Shankar Prasad (ed.), *Time in Indian Philosophy*, Delhi: Indian Books, 1992.
75. Anthony F. Aveni, *Empires of Time: Calendars, Clocks, and Cultures*, Japan: Kodansha International Ltd., 1995.
76. Mircea Eliade, *The Myth of the Eternal Return*, Princeton, NJ: Princeton University Press, 1971.
77. Barbara Adam, *Timewatch: The Social Analysis of Time*, Oxford: Blackwell Publishers, 1995.
78. Michael Batty and Sam Cole (eds.), 'Time and Space: Geographic Perspective on the Future', special issue of *Futures*, 29(4–5), May–June 1997.
79. ibn Khaldun, *The Muqaddimah: An Introduction to History*, translated by Franz Rosenthal, Princeton, NJ: Princeton University Press, 1967.
80. al-Ghazzali, *The Book of Knowledge*, translated by Nabih Amin Faris, Lahore: Ashraf, 1966.
81. Confucius, *The Analects*, translated by D. C. Lau, London: Penguin, 1979.
82. *The Book of Menicus*, translated by D. C. Lau, London: Penguin, 1970.

83. Kalidasa, *Sakuntala*, translated by Michael Coulson, London: The Folio Society, 1992
84. Krishan Dutta and Andrew Robinson (eds.), *Rabindranath Tagore: An Anthology*, London: Picador, 1997.
85. Mohandas K. Gandhi, *Autobiography: The Story of Experiments with Truth*, New York: Dover, 1983.
86. Sei Shonagon, *The Pillow Book of Sei Shonagon*, translated by Ivan Morris, 2 vols., New York: Columbia University Press, 1967.
87. Yoshida no Kenko, *Essays on Idleness*, translated by Donald Keene, New York: Columbia University Press, 1967.
88. David Marshall Lang (ed.), *A Guide to Eastern Literatures*, London: Weidenfeld and Nicolson, 1971.
89. Ian P. McGreal (ed.), *Great Thinkers of the Eastern World*, New York: HarperCollins, 1995.
90. Helaine Selin (ed.), *Encyclopaedia of the History of Science, Technology, and Medicine in Non-Western Cultures*, Dordrecht: Kluwer, 1997.
91. Kwame Anthony Appiah and Henry Louis Gates, Jr. (eds.), *The Dictionary of Global Cultures*, London: Penguin, 1997.
92. George Thomas Kurian and Graham T. T. Molitor (eds.), *Encyclopaedia of the Future*, 2 vols., New York: Macmillan, 1996.

Index